BESTSELLING BOOK SERIES

Ham Radio For Dummies

Cheat Sheet

Wiley, the Wiley Publishing logo, For Dummies, the Dummies Man logo, the For Dummies Bestselling Book Series logo and all related trade dress are trademarks or registered trademarks of John Wiley & Sons, Inc. and/or its affiliates. All other trademarks are property of their respective owners.

Your Emergency Frequency Reference

Fill in the blanks and use this section as a handy reference for your emergency frequencies. Keep it accessible and carry it in your purse or wallet.

Net or Activity	Name	Frequency	Times
Local Emergency Net			
Local Emergency Net			
State Emergency Net			
Regional Service Net			
Regional Service Net			

Technician Class Frequency Privileges

Band	Frequencies (in MHz)	Mode	Notes
80 Meters	3.675 – 3.725	CW	5 wpm code required
40 Meters	7.100 – 7.150	CW	5 wpm code required
15 Meters	21.100 – 21.200	CW	5 wpm code required
10 Meters	28.100 – 28.300 28.300 – 28.500	CW, RTTY, Data CW, Phone, Image	5 wpm code required
Above 50 MHz			All amateur privileges

General Class Frequency Privileges

Band	Frequencies (in MHz)	Mode
160, 60, 30 Meters	All amateur privileges	
80 Meters	3.525 – 3.750 3.850 – 4.000	CW, RTTY, Data CW, Phone, Image
40 Meters	7.025 – 7.150 7.225 – 7.300	CW, RTTY, Data CW, Phone, Image
20 Meters	14.025 – 14.150 14.225 – 14.350	CW, RTTY, Data CW, Phone, Image
15 Meters	21.025 – 21.200 21.300 – 21.450	CW, RTTY, Data CW, Phone, Image
17, 12, 10 Meters	All amateur privileges	
Above 50 MHz	All amateur privileges	

A complete chart of the U.S. frequency and mode privileges for all license classes available is at www.remote.arrl.org/FandES/field/regulations/bands.html.

For Dummies: Bestselling Book Series for Beginners

BESTSELLING
BOOK SERIES

Ham Radio For Dummies®

Cheat Sheet

Common Repeater Channel Spacings and Offsets

Band	Output Frequencies of Each Group (in MHz)	Offset from Output to Input Frequency
6-meters	51.62 – 51.98 52.5 – 52.98 53.5 – 53.98	– 500 kHz
2-meters (there is a and 15kHz channel spacing)	145.2 – 145.5 146.61 – 147.00 147.00 – 147.39	– 600 kHz – 600 kHz + 600 kHz
220 MHz	223.85 – 224.98	– 1.6 MHz
440 MHz (local options determine whether inputs are above or below outputs)	442 – 445 (California repeaters start at 440 MHz) 447 – 450	+ 5 MHz – 5 MHz
1296 MHz	1282 – 1288	– 12 MHz

Ten Handy Ham Radio Web Sites

URL	Organization and Use
www.arrl.org	American Radio Relay League (ARRL): Many useful regulatory, educational, operating, and technical items and links
www.ac6v.com	General-interest Web site with many links on all phases of ham radio
www.qrz.com	Call sign lookup service and general-interest ham radio portal
www.eham.net	News, articles, equipment swap-and-shop, product reviews, mailing lists
www.hfradio.org/ propagation.html	Real-time information on propagation and solar data
www.tapr.org	Tucson Amateur Packet Radio: Information on all popular digital data modes
www.amsat.org	Radio Amateur Satellite Corp: Main site for information on amateur satellites
www.hornucopia. com/contestcal	Contest calendar and log due dates
www.arrl.org/tis/	ARRL Technical Information Service: Technical articles, literature, and vendor searches
DXsummit.com	Worldwide DX spotting network

Common Q Signals

Q Signal	Meaning
QRL	Is the frequency busy? The frequency is busy. Please do not interfere.
QRM	Abbreviation for interference from other signals.
QRN	Abbreviation for interference from natural or man-made static.
QRO	Shall I increase power? Increase power.
QRP	Shall I decrease power? Decrease power.
QRQ	Shall I send faster? Send faster (__WPM).
QRS	Shall I send more slowly? Send more slowly (__WPM).
QRT	Shall I stop sending? Stop sending.
QRU	Have you anything more for me? I have nothing more for you.
QRV	Are you ready? I am ready.
QRX	Standby.
QRZ	Who is calling me?
QSB	Abbreviation for signal fading.
QSL	Received and understood.
QSO	Abbreviation for a contact.
QST	General call preceding a message addressed to all amateurs.
QSX	I am listening on ___ kHz.
QSY	Change to transmission on another frequency (or to ___ kHz).
QTH	What is your location? My location is ____.

Item 5987-7.
For more information about Wiley Publishing,
call 1-800-762-2974.

For Dummies: Bestselling Book Series for Beginners

Ham Radio
FOR
DUMMIES®

Ham Radio
FOR
DUMMIES®

by Ward Silver

WILEY

Wiley Publishing, Inc.

Ham Radio For Dummies®

Published by
Wiley Publishing, Inc.
111 River Street
Hoboken, NJ 07030-5774

Copyright © 2004 by Wiley Publishing, Inc., Indianapolis, Indiana

Published by Wiley Publishing, Inc., Indianapolis, Indiana

Published simultaneously in Canada

For general information on our other products and services or to obtain technical support, please contact our Customer Care Department within the U.S. at 800-762-2974, outside the U.S. at 317-572-3993, or fax 317-572-4002.

Wiley also publishes its books in a variety of electronic formats. Some content that appears in print may not be available in electronic books.

Library of Congress Control Number: 2004101969

ISBN: 0-7645-5987-7

Manufactured in the United States of America

10 9 8 7 6 5 4 3 2

1B/QV/QU/QU/IN

WILEY

About the Author

Ward Silver NØAX has been a licensed ham since 1972 at the age of 17. Ward's experiences in ham radio contributed greatly to a 20-year career in broadcasting and as an electrical engineer and programmer, developing instrumentation and medical electronics. In 2000, he turned to teaching and writing as a second career. He is currently adjunct faculty with Seattle University's Electrical and Computer Engineering department, concentrating on laboratory instruction. You can find his monthly columns and articles in QST magazine and in the biweekly e-mail newsletter, "The Contester's Rate Sheet." He is the author of the ARRL's online course, "Antenna Design and Construction." His ham radio interests include multi-operator and low-power (QRP) contesting, DX-ing, and antenna design. He is the author of "NØAX's Radio Puzzler" (a collection of quizzes and puzzles), and co-author (with K7LXC) of "HF Tribander Performance - Test Methods & Results" and "HF Vertical Performance - Test Methods & Results." He is the winner of the 2003 Bill Orr Technical Writing Award.

Dedication

This book is dedicated to my all-ham family — my wife Nancy W7FIR and sons Webster KD7FYX and Lowell KD7DQO — who frequently see little of Dad except the back of his head as he hammers away at the keyboard creating another article or chapter. And dedicated as well to my own ham radio Elmers: Bill KJ7PC, Randy WB9FSL, Jerry WAØACF, and Danny K7SS who helped me turn "Can I?" into "I did!" Thanks, one and all!

Author's Acknowledgments

I would like to gratefully acknowledge the assistance of the American Radio Relay League for providing a number of excellent graphics and for its extensive Web site that provides so much value to many hams around the world. I also thank the administrators of eHam.net, QRZ.com, and AC6V.com Web portals for providing many useful links and extensive archives. All these organizations and the creators of the many other Web sites referenced in this book exemplify the best of ham spirit.

Also supporting me in many ways were the members of the Western Washington DX Club W7DX, the Vashon-Maury Island Radio Club W7VMI, and many great friends all over the world that share a deep friendship and appreciation of the magic of radio; you know who you are.

Thank you and Very 73.

Publisher's Acknowledgments

We're proud of this book; please send us your comments through our online registration form located at www.dummies.com/register/.

Some of the people who helped bring this book to market include the following:

Acquisitions, Editorial, and Media Development

Project Editor: Christopher Morris

Acquisitions Editor: Melody Layne

Technical Editor: Kirk A. Kleinschmidt

Editorial Manager: Kevin Kirschner

Media Development Supervisor: Richard Graves

Editorial Assistant: Amanda Foxworth

Cartoons: Rich Tennant (www.the5thwave.com)

Production

Project Coordinator: Courtney MacIntyre

Layout and Graphics: Amanda Carter, Andrea Dahl, Lauren Goddard, Denny Hager, Joyce Haughey, Michael Kruzil, Jacque Schneider, Julie Trippetti, Mary Gillot Virgin

Proofreaders: Carl Pierce, Brian H. Walls, TECHBOOKS Production Services

Indexer: TECHBOOKS Production Services

Publishing and Editorial for Technology Dummies

    Richard Swadley, Vice President and Executive Group Publisher

    Andy Cummings, Vice President and Publisher

    Mary C. Corder, Editorial Director

Publishing for Consumer Dummies

    Diane Graves Steele, Vice President and Publisher

    Joyce Pepple, Acquisitions Director

Composition Services

    Gerry Fahey, Vice President of Production Services

    Debbie Stailey, Director of Composition Services

Contents at a Glance

Table of Contents

Introduction

. . . .

*A*mateur or ham radio has been around for nearly a century. In that time, it's grown, branched, morphed, and amplified itself into a worldwide community of licensed hams tickling the airwaves with every conceivable means of communications technology. Its practitioners range in age from preschoolers to septuagenarians. Ham radio's siren call attracts those who have never held a microphone as well as deep technical experts who grew up with a soldering iron and computer.

You may have come across ham radio in any number of ways: in movies such as *Frequency* or *Contact,* in books (the comic book hero TinTin is a ham), from seeing them in action performing emergency communications services, or maybe from a friend or relative who enjoys the hobby. Interestingly enough, ham radio has room for all of these activities. Yes, even a mad scientist or two are in the ham ranks. Most, however, are just like you.

The storied vision finds the ham bent over a glowing radio, surrounded by all manner of electronic gadgets and flicking meters, tapping out messages on a telegraph key or speaking into a large, round, silvery microphone. Today's ham has many more options than that to try! While the traditional shortwave bands are still crowded with ham signals hopping around the planet, hams now transmit data and pictures through the airwaves, use the Internet, lasers, and microwave transmitters, and travel to unusual places high and low to make contact.

Simply stated, ham radio provides the broadest and most powerful wireless communications capability available to any private citizen anywhere in the world. Because the world's citizens are craving ever-closer contact, ham radio is attracting attention from people like you. The hobby has never had more to offer and shows no sign of slowing its expansion into new wireless technologies. Did I say wireless? Think Extreme Wireless!

About This Book

I wrote *Ham Radio For Dummies* for beginning hams. If you just became interested in ham radio, you find plenty of information here to explain what the hobby is all about and how to go about joining the fun by discovering the basics and getting a license.

If you already received your license, congratulations! This book helps you change from a listener to a doer. Any new hobby, particularly a technical one, can be intimidating to newcomers. By keeping *Ham Radio For Dummies* handy in your radio shack, getting your radio on the air and making contacts is easy. I cover the basics of getting a station put together properly and the fundamentals of on-the-air behavior. Use this book as your personal radio buddy and soon you'll sound like a pro!

Conventions Used in This Book

To make the reading experience as clear and uncluttered as possible, I use a consistent presentation style. Here are the conventions:

- ✔ *Italics* note a new or important term.
- ✔ Web site URLs (addresses) are indicated with a `monospace` font.

Foolish Assumptions

In writing this book I made some assumptions about you. You don't have to know a single thing about ham radio or its technology to enjoy *Ham Radio For Dummies.* And you definitely don't need to be an electrical engineer to enjoy this book.

But I ask two things of you: You have an interest in ham radio and that you know how to use a computer well enough to surf the Web. Due to the complicated and extensive nature of ham radio, I couldn't include everything in this book (otherwise, you wouldn't be able to lift it). However, I steer you in the direction of additional resources to check out, including Web sites.

How This Book Is Organized

Ham Radio For Dummies has two major sections. Parts I and II are for readers getting interested in ham radio and preparing to get a license. Parts III and IV explain how to set up a station, get on the air, and make contact with other hams.

The *Ham Radio For Dummies* Web site, at www.dummies.com/go/hamradio, offers a list of Web resources and more information on the technical aspects of this wonderful hobby.

Part 1: What Is Ham Radio All About?

If you don't know much about ham radio, start reading this part. You get the ham radio big picture. Then I send you on a tour of the various radio technologies necessary to get you on the air. I round out this first part with an overview of the ham community: clubs and organizations.

Part II: Wading through the Licensing Process

The four chapters in Part II take you every step of the way through the process of getting a ham radio license. I explain the overall licensing system, including the types of licenses and the volunteers that administer the exams. Then I move on to studying, including Morse code, for your exam. Finally, I discuss the actual exam process so you know what to expect when the time comes. Part II concludes with what to do after you pass your test.

Part III: Hamming It Up

The sky is the limit, but first you have to learn to fly. Part III is where you delve into the fundamentals of ham radio operating. Then you get down to the brass tacks of basic operation, including how to make that elusive first contact! I cover the different kinds of casual operating and then get into some of the popular specialties of the hobby, including public service and emergency communications.

Part IV: Building and Operating a Station That Works

Part IV takes you through the basics of setting up and using a suitable station. I cover the different kinds of ham radio equipment and how to acquire what you need to get your station up and running. Ham radios take a little maintenance and troubleshooting, and I devote a chapter to these topics.

Part V: The Part of Tens

Familiar to all *For Dummies* readers, this part is where the accumulated knowledge and wisdom of the ages is boiled down into several condensed lists. I cover the tips and secrets of ham radio along with general guiding principles for successful ham radio operation.

Part VI: Appendixes

If you come across an unfamiliar term, turn to the glossary. I have also collected a long list of excellent references — both online and off — for you to find and use.

Icons Used in This Book

Ham Radio For Dummies includes icons that point out special information. Here are the icons I use and what they mean:

This icon points out easier, or shorter, ways of doing something.

This icon signals when I show my techy side. If you don't want to know the technical details, skip paragraphs marked with this icon.

Whenever I could think of a common problem or "oops," you see this icon. Before you become experienced, getting hung up on some of these little things is easy.

This icon lets you know that some safety rules or performance issues are associated with the topic of discussion. Watch for this icon to avoid common gotchas.

This icon points out information that you need to remember to enjoy your ham radio experience.

Where to Go from Here

If you're not yet a ham, I highly recommend that you find your most comfortable chair and read through Parts I and II. You discover the basics about ham radio and solidify your interest. For the licensed ham, browse through Parts III and IV to find the topics that most interest you. Take a look through the appendixes, as well, to find out what information is secreted away back there for when you need it in a hurry. And for more technical material — and a list of Web resources — try the *Ham Radio For Dummies* Web site. For all my readers, welcome to *Ham Radio For Dummies* and I hope to meet you on the air some day!

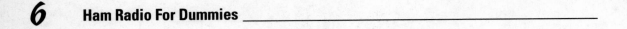

Part I
What Is Ham Radio All About?

The 5th Wave   By Rich Tennant

In this part . . .

*G*et ready to dive into the details of a terrific hobby and service. In this part, you take a quick peek at Ham Radio Land, formally known as the Amateur Radio Service. I start by describing what ham radio is like today — you will be surprised at its breadth and technical sophistication. You get a peek at the many ways in which hams actually communicate and why. Then I show you what most hams have in an actual radio shack.

Ham radio puts a technical face on many of its facets, so you get a quick refresher or introduction to the technology of radio. Then I explain the basic gadgets of ham radio, explain radio waves, and touch on how radio depends on the natural environment.

You certainly don't have to know everything about ham radio by yourself because a lot of other hams are out there to lend a hand. In fact, helping others is one of the long-standing traditions of ham radio, so I show you how to find a local club and join a ham radio organization: you can find a large or small club, or one that caters to your general interest and special purpose. They're just what you need to get in-depth knowledge about your own interests. You also get an introduction to hamfests and the many ham radio gatherings. You are not alone!

Chapter 1

Getting Acquainted with Ham Radio

*H*am radio invokes a wide range of visions. Maybe you have a mental image of a ham radio operator (or *ham*) from a movie or newspaper article. But hams are a varied lot — from go-getter emergency communicators to casual chatters to workshop tinkerers. Everyone has a place, and you do, too.

Hams use all sorts of radios and antennas on a wide variety of frequencies to communicate with other hams across town and around the world. They use ham radio for personal enjoyment, for keeping in touch with friends and family, for emergency communications, and for experimenting with radios and radio equipment. They communicate using microphones, telegraph or Morse keys, computers, cameras, lasers, and even their own satellites.

Hams meet on the air and in person. Ham radio clubs and organizations are devoted to every conceivable purpose. They have special ham radio flea markets and host conventions, large and small. Hams as young as six years old and centenarians have been hams since before ham radio licenses. Some have a technical background, but most do not. One thing all these diverse individuals do have, however, is an interest in radio that can express itself in many different ways.

Ham: Not just for sandwiches anymore

Everyone wants to know the meaning of the word "ham," but as with many slang words, the origin is murky. Theories abound, of course, ranging from the initials of an early radio club's operators to the use of a meat tin as a natural sound amplifier. Out of the many possibilities, this theory condensed from the American Radio Relay League's (ARRL) Web site seems the most believable:

"*Ham:* a poor operator" was used in telegraphy even before radio. The first wireless operators were landline telegraphers who brought with them their language and much of the tradition of their older profession. Government stations, ships, coastal stations, and the increasingly numerous amateur operators all competed for signal supremacy in each other's receivers. Many of the amateur stations were very powerful and could effectively jam all the other operators in the area. When this logjam happened, frustrated commercial operators would send the message "THOSE HAMS ARE JAMMING YOU." Amateurs, possibly unfamiliar with the real meaning of the term, picked it up and wore it with pride. As the years advanced, the original meaning has completely disappeared.

Tuning In Ham Radio Today

Hams enjoy three different aspects of ham radio — the technology, operating, and social points of view. Your interest in the hobby may be technical; you may want to use ham radio for a specific purpose; or you may just want to join the fun. All are perfectly valid reasons for getting a ham radio license.

Using electronics and technology

Ham radio is full of electronics and technology (see Chapter 2). To start with, transmitting and receiving radio signals is a very electronics-intensive endeavor. After you open the hood on ham radio, you're exposed to everything from basic direct-current electronics to cutting-edge radio-frequency techniques. Everything from analog electronics to the very latest in digital signal processing and computing is available in ham radio. I've been in the hobby for more than 30 years and I've never met anyone who is an expert on it all.

You may choose to design and build your own equipment or assemble a station from factory-built components, just like an audiophile might do. All that you need for either path is widely available in stores and on the Web. Hams delight in a do-it-yourself ethic known as *homebrewing* and help each other out to build and maintain their stations.

Hams also develop their own software and use the Internet along with radios to create novel hybrid systems. Hams developed packet radio by adapting data transmission protocols used over computer networks to amateur radio links. Packet radio is now widely used in many commercial applications. By combining GPS radiolocation technology with the Web and amateur mobile radios, the Automatic Position Reporting System (APRS) was developed and is now widely used. More information about these neat systems is contained in Parts III and IV.

Voice and Morse code communications are still the most popular technologies by which hams talk to each other, but computer-based digital operation is gaining fast. The most common home station configuration today is a hybrid of the computer and radio. Some of the newer radios are exploring software-defined radio (SDR) technology that allows reconfiguration of the circuitry that processes radio signals under software control.

Along with the equipment and computers, hams are students of antennas and *propagation,* which is the means by which radio signals bounce around from place to place. Hams take an interest in solar cycles, sunspots, and how they affect the Earth's ionosphere. For hams, weather takes on a whole new importance, generating static or fronts along which radio signals can sometimes travel long distances. Antennas, with which signals are launched to take advantage of all this propagation, provide a fertile universe for the station builder and experimenter.

Antenna experimentation is a hotbed of activity for hams. New designs are created every day and hams have contributed many advances and refinements to the antenna designer's art. Antenna systems range from small patches of printed circuit board material to multiple towers festooned with large rotating arrays. All you need is some wire, a feedline, and a soldering iron.

Hams also use radio technology in support of hobbies such as radio control (R/C), model rocketry, and meteorology. Hams have special frequencies for R/C operation in the 6-meter band, away from the crowded unlicensed R/C frequencies. Miniature ham radio video transmitters are frequently flown in model aircraft, rockets, and balloons, beaming back pictures from heights of hundreds and thousands of feet. Ham radio data links are also used in support of astronomy, aviation, auto racing and rallies, and many other pastimes.

Whatever part of electronic and computing technology you most enjoy, it's all used in ham radio somewhere . . . and sometimes all at once!

Operating a ham radio: Making contacts

If you were to tune a radio across the ham bands, what would you hear hams doing? Contacts run the range from simple conversation to on-the-air meetings to contesting (recording the highest number of contacts).

Ragchews

By far the most common type of activity for hams is just engaging in conversation, which is called *chewing the rag;* such contacts are called *ragchews.* Ragchews take place between continents or across town. You don't have to know another ham to have a great ragchew — ham radio is a very friendly hobby with little class snobbery or distinctions. Just make contact and start talking! Find out more about ragchews in Chapter 9.

Nets

Nets (an abbreviation for networks) are organized on-the-air meetings scheduled for hams with a similar interest or purpose. Some of the nets you can find are

- **Traffic nets:** These are part of the North American system that moves text messages or *traffic* via ham radio. Operators meet to exchange or *relay* messages, sometimes handling dozens in a day. Messages range from the mundane to emergency health-and-welfare.

- **Emergency service nets:** Most of the time, these nets just meet for training and practice. When disasters or other emergencies strike, hams organize around these nets and provide crucial communications into and out of the stricken areas until normal links are restored.

- **Technical Service:** These nets are like radio call-in programs in which stations call with specific questions or problems. The net control station may help, but more frequently, one of the listening stations contributes the answer. Many are designed specifically to assist new hams.

- **ALE Mailboxes and Bulletin Boards:** If you could listen to Internet systems make contact and exchange data, this is what they'd sound like. Instead of transmitting 1s and 0s as voltages on wires, hams use tones. *ALE* stands for *Automatic Link Establishment* and means that a computer system is monitoring a frequency all the time so that others can connect to it and send or retrieve messages. Sailors and other travelers use ham radio where the Internet isn't available.

- **Swap Nets:** In between the in-person hamfests and flea markets, in many areas a weekly swap net allows hams to list items for sale or things they need. A net control station moderates the process and business is generally conducted over the phone once the parties have been put in contact with each other.

It's better than "masticating the towel"

Unlike "ham," the origins of "ragchew" are fairly clear. The phrase "chewing the rag" is well known back to the late Middle Ages. "Chew" was slang for "talk," and "rag" is derived from "fat," or is a reference to the tongue. "Chewing the rag" thus became a phrase referring to conversation, frequently while sitting around a meal. Hams picked up that usage from telegraphers, and because most of ham radio is, in fact, conversations, it has been a part of radio from its earliest days.

DX-ing, contests, and awards

DX stands for *distance* and the lure of making contacts ever-farther from home has always been a part of ham radio. Hams compete to contact faraway stations and to log contacts with every country. They enjoy contacting islands and making personal friends in a foreign country. When conditions are right and the band is full of foreign accents, succumbing to the lure of DX is easy!

Ham radio's version of rugby, contests are events in which the point is to make as many contacts as possible, sometimes thousands, during the contest time period, by sending and receiving short messages. These exchanges are related to the purpose of the contest — to contact a specific area, use a certain band, find a special station, or just contact everybody.

Along with contests, thousands of special-event stations and awards are available for various operating accomplishments, such as contacting different countries or states. For example, in December 2003, the station W4B was set up at Kitty Hawk, North Carolina, and operated during the centennial of the Wright Brothers' first flight.

DX-ing, contests, and awards are closely related, and if you enjoy the thrill of the chase, go to Chapter 11 to find out more about all of these activities.

Joining the ham radio community

Because of their numbers and reliance on uncomplicated infrastructure, hams are able to bounce back quickly when a natural disaster or other emergency makes communications over normal channels impossible. Hams organize themselves into local and regional teams that practice responding to a variety of emergency needs, working to support public safety agencies such as police and fire departments.

Is it hurricane season? Every fall in North America, ham emergency teams gear up for these potentially devastating storms. Hams staff an amateur station at the National Hurricane Center in Florida (www.fiu.edu/orgs/w4ehw/) and keep the Hurricane Watch Net busy on 14.325 MHz (www.hwn.org/). After the storm, hams are the first voices heard from the affected areas with many more standing by to relay their messages and information.

After the September 11, 2001, terrorist attacks, hams manned an emergency operations center around the clock for weeks. Government agencies had to focus on coordinating recovery and rescue efforts. The hams were able to handle "health-and-welfare" messages to support the emergency workers in their efforts.

Every June, on the last full weekend, hams across the United States engage in an emergency operations exercise called *Field Day.* It's an opportunity for hams to operate under emergency conditions. An amateur emergency team or station probably is operating in your town or county.

Hams provide assistance for more than emergencies. Wherever there is a parade, festival, marathon, or other opportunity to provide communications services, you may find ham radio operators helping out. In fact, this is great training for emergencies!

A particularly beneficial relationship exists between ham radio and *philately,* or stamp collecting. Hams routinely exchange postcards called *QSLs* with their call signs, information about their stations, and often colorful graphics or photos. Stamp collecting hams combine the exchange of QSLs with collecting by sending the cards around the world with local colorful stamps or special postmarks. Foreign hams return the favor with a stamp of their own. The cheerful greeting of those red-and-blue airmail envelopes from an exotic location is a special treat!

Hams like to meet in person as well as on the radio. Membership in at least one radio club is a part of nearly every ham's life. In fact, in some countries, you're required to be a member of a club before you can even get a license. Chapter 3 shows you how to find and join clubs — they're great sources of information and assistance for new hams.

The two other popular ham gatherings are *hamfests* and conventions. A hamfest is a ham radio flea market where hams bring their electronic treasures for sale or trade. Some are small, parking-lot-size get-togethers on a Saturday morning while others attract thousands of hams from all over the world and last for days. These are more like the conventions hams hold with a variety of themes from public service to DX and low-power operating. Hams travel all over the world to attend conventions and meet friends known only as a voice and a call sign over the crackling radio waves.

Roaming the World of Ham Radio

Although the United States has a large population of hams, it by no means represents the majority. The amateur population in Europe is growing by leaps and bounds, and Japan has an even larger amateur population. With more than 3 million hams worldwide, very few countries are without an amateur.

Hams are required to have a license, no matter where they operate. The international agency that manages radio activity is the International Telecommunication Union, or ITU (www.itu.int/home/). Each member country is required to have its own government agency that controls licensing inside its borders. In the United States, hams are part of the Amateur Radio Service, which is regulated and licensed by the Federal Communications Commission (FCC). Outside the United States, Amateur Radio is governed by similar rules and regulations.

Amateur Radio licenses in America are granted by the FCC, but the tests are administered by other hams acting as *volunteer examiners,* or *VEs.* I discuss VEs in detail in Chapter 4. Classes and testing programs are often available through local clubs.

Since the adoption of international licensing regulations, hams operate from many different countries with a minimum of paperwork. For example, a ham from a country that is a party to the international license recognition agreement known as CEPT can use his or her home license to operate from within any other CEPT country. The ARRL has gathered a lot of useful material about international operating on its Web site at www.arrl.org/FandES/field/regulations/io.

Hams across the world

Where are the hams and how many are around this big world? Over 3 million populate the amateur bands, although not all are equally active. As of 2000, the International Amateur Radio Union (IARU) counted 195 different countries with a national radio society. The growing countries of the Pacific Rim have substantial amateur populations. Europe, Africa, and Russia total 442,193. The Americas total nearly 1 million with 830,492. Asia and the Pacific countries have the most at 1,714,087. Amateur numbers are showing moderate growth in North America and strong growth in Asia and Europe. Tune the bands on a busy weekend and you'll see what I mean!

Because radio signals know no boundaries, hams have always been in touch across the political borders. Even during the Cold War, U.S. and Soviet hams made regular contact, fostering long personal friendships and international goodwill. While the Internet makes global communications easy, chatting by voice or Morse code over the airwaves to someone in another country is exciting.

Communicating with Ham Radio

Though you make contacts for different purposes — chatting, emergencies, a net, or to win a contest — most contacts follow the same structure.

After you get a response from your call or respond to someone else calling, you exchange names, information about who you are, and the quality of your signal to gauge conditions. If you're chatting, you can talk about how you constructed your station, what you do for a living, your family, and your job.

Except for the fact that you take turns transmitting and information is converted to radio waves that bounce off the upper atmosphere, contacts are just like talking to someone that you meet at a party or convention. You can hold the same conversation by voice, using Morse code, or typing from keyboard to keyboard using computers as intermediaries to the radios. You won't find great purpose behind the average contact except a desire to meet another ham and see where your radio signal can be heard.

A frequent question asked about ham radio i: ╵o know where to tune for a certain station?" and the answer is usually, ╵dio operators don't have specific frequency assignments o news is that ham radio has an unparalleled communications under continually changing that making contact with one specific static know on what frequency to call them. How around the latter problem with the result be adaptive communications service.

Building a Ham Radio SF

The term *radio shack,* for me, conjures vis tist's lab than a modern ham station. But ; you keep your radio and ham equipment. jumping meters, and two-handed control

Shacks that aren't Lakers

Where did the phrase, radio shack, come from? Back in the early days of radio, the equipment was highly experimental and all home-built, requiring a nearby workshop. In addition, the first transmitters used a noisy spark to generate radio waves. The voltages were high and the equipment somewhat of a mess, so the radio hobbyists often found themselves banished from the house proper. Thus, many early stations were built in a garage or tool shed. The term "shack" was only natural and carries through today as a description of the state of order and cleanliness found in many a ham's lair.

For some hams, the entire shack consists of a hand-held radio or two. Other hams operate on the go in a vehicle. Cars make perfectly good shacks, but most hams have a spot somewhere at home they claim for a ham radio. Here's what you can find in a ham shack:

- **The rig:** The offspring of the separate receiver and transmitter of yore, the modern radio or *rig* combines both in a single, compact package about the size of a large DVD player. Like its ancestors, a large tuning knob controls the frequency. Unlike them, state-of-the-art digital displays replace the dials and meters.

- **Computer:** A majority of hams today have at least one computer in the shack. Computers now control many radio functions (including keeping records). Using digital data communications simply wouldn't be possible without one. Some hams use more than one computer at a time.

- **Mobile/base rig:** For operating on the local repeater stations, hams may use a hand-held radio, but in the shack a more capable radio is used. These units are about the size of a good-sized hardcover book and you can use them as either a mobile or base rig.

- **Microphones, keys, and headphones:** Depending on the shack owner's preferences, you see a couple (or more!) of these important gadgets, the radio's true user interface. Mikes and keys range from imposing and chrome-plated to miniaturized and hidden. The old Bakelite headphones or *cans* are also a distant memory (good — they hurt my ears!), replaced with lightweight and comfortable, hi-fi quality designs.

- **Antennas:** In the shack, you find switches and controllers for antennas that live outside the shack. Outside, a ham shack tends to sprout antennas ranging from vertical whips the size of a pencil to wire antennas stretched through the trees and on up to super-sized directional beams held high in the air on steel towers. See Chapter 12 for more info on antennas.

✔ **Cables and feedlines:** Look behind, around, or under any piece of shack equipment and you find wires. Lots of them. The radio signals pipe through fat, black round cables called *coaxial* (coax). You're probably familiar with audio cables from stereo equipment. Power is supplied by colored wires not terribly different in size from house wiring. I cover cables and feedlines in more detail in Chapter 12.

Although you perform many of the same functions as the hams from the nineteenth century, the modern shack is as far removed from the home-brewed breadboards in the backyard shed as a late model sedan is from a Model T. You can see examples of several different shacks, including mine, in Chapter 13.

Chapter 2

Getting a Handle on Ham Radio Technology

*H*am radio covers a lot of technological territory. And to be successful in ham radio, you need to have a general understanding of the technology that makes up ham radio. In this chapter, I cover the most common terms and ideas that form the foundation of ham radio.

Fundamentals of Radio Waves

Understanding ham radio (or any radio) is impossible without also having a general understanding of the purpose of radio — to send and receive information by using *radio waves*.

Radio waves are just another form of light and travel at the same speed: 186,000 miles per second. Radio waves can get to the moon and back in 2½ seconds or circle the Earth in ½ of a second.

An electric field and a magnetic field carry the energy of a radio wave. These fields affect charged particles, such as the electrons in a wire, and make them move. Electrons move in specific ways: They move parallel to electric fields and in circular motions in response to magnetic fields. These moving electrons (that is, *current*) also create moving electric and magnetic fields (in other words, radio waves) and vice versa.

Transmitters cause electrons to move so that they, in turn, create the moving fields of radio waves. Antennas are just structures for electrons to move in to create radio waves. The electrons in an antenna also move in response to radio waves launched by other antennas. Receivers then detect the electron motion caused by the incoming radio waves. The energy is just transferred from electrons to radio waves and back to electrons at the other station.

Frequency and wavelength

The radio wave-electron relationship has a wrinkle. The fields of the radio wave aren't just one strength all the time — they vary between a positive and a negative value, or *oscillate,* like a vibrating string moves above and below its stationary position. The time a field's strength takes to go through one complete set of values is called a *cycle.* The number of cycles in one second is the *frequency* of the wave, measured in hertz (abbreviated Hz).

One other wrinkle — the wave is also moving at the speed of light, which is constant. If you could watch the wave oscillate as it moved, you would see that the wave always moves the same distance (one *wavelength*) in one cycle. The higher the wave's frequency, the quicker a cycle completes, and the less time it has to move during one cycle. High frequency waves have short wavelengths and low frequency waves have long wavelengths.

If you know a radio wave's frequency, you can figure out the wavelength because the speed of light is always the same. Here's how:

Wavelength = Speed of light / Frequency of the wave

Wavelength in meters = 300,000,000 / Frequency in hertz

Similarly, if you know how far the wave moves in one cycle (the wavelength), you also know how fast it oscillates because the speed of light is fixed.

Frequency in hertz = 300,000,000 / Wavelength in meters

Frequency is abbreviated as f, the speed of light as c, and wavelength as the Greek letter lambda, λ, leading to the following simple equations:

$f = c / \lambda$ and $\lambda = c / f$

Radio waves oscillate at frequencies between a few hundred kilohertz, or kHz (kilo is the metric abbreviation meaning 1,000), up to 1,000 gigahertz, or GHz (giga is the metric abbreviation meaning 1 billion). They have corresponding wavelengths from hundreds of meters at the low frequencies to a fraction of a millimeter (mm) at the high frequencies.

The most convenient two units to use in thinking of radio wave frequency (RF) and wavelength are megahertz (MHz; mega means 1 million) and meters (m). The equation describing the relationship is much simpler when you use MHz and m:

$$f = 300 / \lambda \text{ in m and } \lambda = 300 / f \text{ in MHz}$$

If you are not comfortable with memorizing equations, an easy way to convert frequency and wavelength is to memorize just one combination: for example, 300 MHz and 1 meter or 10 meters and 30 MHz. Then use factors of ten to move in either direction, making frequency larger and wavelength smaller as you go.

The radio spectrum

The range, or *spectrum,* of radio waves covers a lot. Tuning a radio receiver to different frequencies, you hear radio waves carrying all kinds of different information. These frequencies are called *signals* and are grouped by the type of information they carry into different ranges of frequencies, called *bands.* For example, AM broadcast band stations are at frequencies between 550 and 1700 kHz (550,000 and 1,700,000 hertz or 0.55 and 1.7 MHz). Bands help you find the signals, without having to hunt over a wide range.

The different users of the radio spectrum are called *services,* such as the broadcasting service or Amateur Radio Service. Each service gets a certain amount of spectrum to use, called a *frequency allocation.* Amateur Radio, or ham radio, has quite a number of allocations sprinkled throughout the radio spectrum. I get into the exact locations for the ham radio bands in Chapter 8.

Radio waves at different frequencies act differently in the way they travel and require different techniques to transmit and receive. Because waves of similar frequencies tend to have similar properties, the radio spectrum is divided into four segments:

- **Shortwave or High-Frequency (HF):** Frequencies below 30 MHz. Includes AM broadcasting, ten different ham radio bands, ship-to-shore and ship-to-ship, military, and Citizens Band.

- **Very High Frequency (VHF):** Frequencies from 30 to 300 MHz. Includes TV channels 2 through 13, FM broadcasting, three ham bands, public safety and commercial mobile radio, and military.

- **Ultra High Frequency (UHF):** Frequencies from 300 MHz to 1 GHz. Includes TV channels 14 and higher, two ham bands, cellular phones, public safety and commercial mobile radio, and military.

- **Microwave:** Frequencies above 1 GHz. Includes GPS, digital wireless telephones, WiFi wireless networking, microwave ovens, eight ham bands, satellite TV, and numerous public, private, and military users.

Because a radio wave has a specific frequency and wavelength, hams use the terms somewhat interchangeably. For instance, the 40-meter and 7 MHz ham bands are the same thing. Check out Chapter 8 for the specifics on the ham bands. I will use both in this book so that you become used to interchanging them as hams are expected to do.

Basic Ham Radio Gadgetry

Although the occasional vacuum tube radio may still be found glowing in an antiques-loving ham's station, today's ham radios are sleek, microprocessor-controlled communications centers. The basic radio is composed of a *receiver* combined with a *transmitter* to make a *transceiver,* called a *rig* by hams. Mobile and handheld radios are called rigs, too. If the rig doesn't use AC line power, use a *power supply* to provide the DC voltage and current. Figure 2-1 shows a typical basic station's components and gadgets.

The radio is connected with a *feedline* to one or more antennas. Two of the most popular are shown in Figure 2-1: A *dipole* is an antenna made from wire and connected to its feedline in the middle. A *beam* antenna sends and receives radio waves better in one direction than others and is often mounted on a tall pole or tower with a rotator that can point it in different directions. *Antenna switches* allow the operator to select one of several antennas. An *antenna tuner* is used between the antenna/feedline combination and the transmitter, like a vehicle's transmission, in order for the transmitter to operate at peak efficiency.

You use headphones and a microphone to communicate using the various methods of transmitting speech. If *Morse code* (or *CW*) is preferred, you can use the traditional *straight key* (an old-fashioned Morse code transmission device), but more commonly you use a *paddle* and *keyer.* A paddle and keyer are much faster than straight keys and require less effort to use. The paddle looks like a pair of straight keys mounted back to back on their sides. The keyer converts the closing and opening of each paddle lever into strings of dots and dashes.

If you use *digital data* for contacts, a computer or other device is required to interpret and generate the on-the-air signal. In this case, disconnect the microphone (and probably headphones) and replace them with connections to the external equipment, as shown in Figure 2-2. A *data interface* passes signals between the radio and computer. For some types of data, a computer can't do necessary processing, so a *multi-protocol controller* is used. The computer talks to the controller using its serial RS-232 (COM) port.

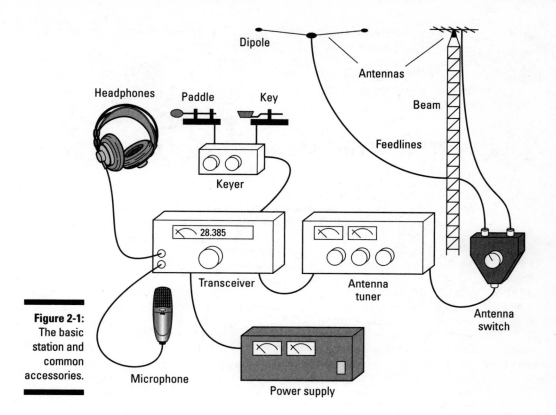

Figure 2-1:
The basic
station and
common
accessories.

Many radios also have an RS-232 interface that allows a computer to control the radio directly. You can find a lot of programs that allow you to change frequency and many other radio settings from the keyboard. Computers can also send and receive Morse code, marrying the hottest twenty-first-century technology with the oldest form of electronic communications!

Miscellaneous gadgets

Aside from the components that make up your actual operating station, quite a number of additional tools and pieces of equipment make up the rest of your gear.

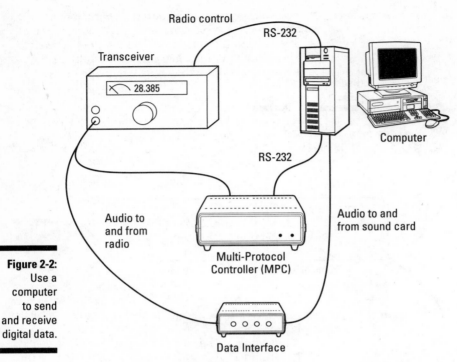

Radio control

RS-232

Transceiver

28.385

Computer

RS-232

Audio to
and from
radio

Audio to and
from sound card

Multi-Protocol
Controller (MPC)

Figure 2-2:
Use a
computer
to send
and receive
digital data.

Data Interface

Feedline measurements

Most radios and antenna tuners have the ability to evaluate the electrical
conditions inside the feedline, measured as the *standing wave ratio (SWR).*
SWR is a ratio of voltages and tells you how much of the power supplied by
the transmitter is getting radiated by the antenna. Most radios have a built-in
meter that shows feedline SWR. Having a stand-alone SWR sensor, called an
SWR meter or an *SWR bridge,* to measure SWR is very handy when working on
antennas or operating in a portable situation. You can also measure feedline
conditions by using a *power meter,* which measures the actual power flowing
back and forth. SWR meters are inexpensive, while power meters are more
accurate. These devices are typically used right at the transmitter output.

Filters

Filters are designed to pass or reject ranges of frequencies. Some filters are
designed to pass or reject only specific frequencies. Filters can be made of
inductors and capacitors, called *discrete components,* or even from sections
of feedline, called *stubs.* Filters come in the following varieties:

✔ **Feedline:** Use *feedline* filters to prevent unwanted signals from getting to the radio from the antenna or vice versa. On transmitted signals, you can use them to ensure that unwanted signals from the transmitter are not radiated, causing interference to others. They also prevent undesired signals from getting to the receiver where they may compromise receiver performance.

✔ **Receiving:** *Receiving* filters are installed inside the radio and are usually made from quartz crystals or from small tuning-fork-like structures. Receiving filters remove all but a single desired signal in a receiver. Filters improve a receiver's selectivity, which refers to its ability to receive the single signal in the presence of many signals.

✔ **Audio:** Use audio filters on the receiver output to provide additional filtering capability, rejecting nearby signals or unwanted noise.

✔ **Notch:** A notch filter works to remove a very narrow range of frequencies, such as a single interfering tone.

Feedlines

Two common types of electrical feedlines connect the antennas to the station and carry RF energy between pieces of equipment. The most common is *coaxial cable,* or just *coax,* so named because it is constructed of a hollow tube surrounding a central wire concentrically. The outer conductor is called the *shield* and also *braid,* if made from fine woven wire. The wire in the middle is called the *center conductor* and is surrounded by insulation that holds it right in the center of the cable. The outer conductor is covered by a plastic coating called the *jacket.* The other kind of feedline is *open-wire,* also called *twin-lead* or *ladder line,* which is made from two parallel wires. The wires may be exposed, only held together with insulating spacers, or plastic insulation may cover them.

Ham Radio on the Air

Aside from the equipment, ham radio technology also extends to making contacts and exchanging information. You use the following technologies when using your ham radio:

✔ **Modulation/Demodulation:** *Modulation* is the process of adding information to a radio signal to transmit over the air. *Demodulation* is the process of recovering information from a received signal. Ham radios use two kinds of modulation: amplitude modulation (AM) and frequency modulation (FM).

✔ **Modes:** Modes are specific types of modulation. You can choose from several modes when transmitting, some of which include voice, data, video, or Morse code.

- **Repeaters:** *Repeaters* are relay stations that listen on one frequency and retransmit what they hear on a different frequency. Because they are often located on tall buildings, towers, or hilltops, they enable hams to use low-power radios to converse over many miles. They can be linked together either by radio or the Internet to extend communication around the world. Repeaters can listen and transmit at the same time, which is called *duplex* operation.

- **Satellites:** Yes, just like the military and commercial services, hams use satellites. Some amateur satellites act as "repeaters in the sky," while others are used as orbiting digital bulletin boards and e-mail servers.

- **Computer software:** Computers have become a big part of ham radio. Initially, they were limited to doing paperwork and making calculations in the shack. Today, they also act as part of the radio, generating and decoding digital data signals, sending Morse code, and controlling the radio's functions. Hams also have constructed radio-linked computer networks and a worldwide system of e-mail servers accessed by radio.

Hams have always been interested in pushing the envelope in applying and developing radio technology — which is one of the fundamental reasons that ham radio exists as a licensed service. Today, that includes creating novel hybrids of radio and other technologies, such as the Internet and GPS radio location. For example, the ARRL High Speed Multimedia Working Group is working to adapt wireless LAN technology to ham radio. The Tucson Amateur Packet Radio (TAPR) organization has many members around the world developing new methods of digital communication. You also find ham radio to be a hotbed of innovation in antenna design and construction. Ham radio is "techie heaven!"

Dealing with Mother Nature

Because radio waves travel through the natural environment on their way to space or to another terrestrial station, its phenomena affect them. Thus hams take a keen interest in propagation.

For local contacts, the radio wave journey along the surface of the Earth is called *ground wave* propagation. Ground wave propagation can support communication up to 100 miles, but varies greatly with the frequency being used.

To make longer range contacts, the radio waves must travel through the atmosphere. At HF and sometimes at VHF, the upper layers of the atmosphere, called the *ionosphere,* reflect the waves back to Earth. The reflection of radio waves is called *sky wave* propagation. Depending on the angle at which the signal is reflected, a sky wave *path* can be as long as 2,000 miles. HF signals often bounce between the Earth's surface and the ionosphere several times so that contacts are made worldwide. At VHF, multiple hops are rare but use other reflecting mechanisms.

Apart from the ionosphere, the atmosphere itself can direct radio waves. *Tropospheric propagation,* or just *tropo,* occurs along weather fronts, temperature inversions, and other large-scale phenomena in the atmosphere. Tropo is common at VHF and UHF frequencies, often supporting contacts over 1,000 miles or more.

Ham radios, CB radios, and cellular phones

Radios abound — enough to boggle your mind. Here are the differences between your ham radio and those other radios:

✔ **Citizens Band (CB):** CB radio uses 40 channels near the 28 MHz ham band. CB radios are low-power and useful for local communications only, although the radio waves sometimes travel long distances. You don't need a license to operate a CB radio. This lightly-regulated service is plagued by illegal operation that diminishes its usefulness.

✔ **Family Radio Service (FRS) and General Mobile Radio Service (GMRS):** These popular radios, such as the Motorola Walkabout models, are designed for short-range communications between family members. Usually hand-held, both types operate with low-power on UHF frequencies. FRS operation is unlicensed, but the higher-power, more capable GMRS radios do require a license.

✔ **Broadcasting:** Although hams are often said to be "broadcasting," this term is incorrect. Hams are barred from doing any one-way broadcasting of programs the way

AM, FM, and TV stations do. Broadcasting without the appropriate license attracts a lot of attention from a certain government agency whose initials are FCC.

✔ **Public safety and commercial mobile radio:** The hand-held and mobile radios used by police, firemen, construction workers, and delivery companies are similar in many ways to VHF and UHF ham radios. In fact, the frequency allocations are so similar that hams often convert surplus equipment. Commercial and public safety radios require a license to operate.

✔ **Cellular and digital wireless telephones:** Obviously, you don't need a license to use a wireless phone, but you can only communicate through a licensed service provider on one of the wireless phone allocations. The older, analog phones operate between 800 and 900 MHz, while the newer digital phones operate near 2 GHz. While the phones are actually little UHF and microwave radios, except for a few models, they can't communicate directly with each other and are completely dependent on the wireless phone network to operate.

Two other VHF- and UHF-reflecting features exist in our atmosphere: the *aurora* and *meteor trails.* When the aurora is strong, it absorbs HF signals, but reflects VHF and UHF signals adding a characteristic rasp or buzz. Hams active on those bands know to point their antennas north to see if the aurora can support an unusual contact. Meteor trails are very hot from the friction of the meteor's passage through the atmosphere. So hot, that the gasses become electrically conductive and reflect signals until they cool. For a few seconds, a radio mirror floats high above the Earth's surface. Meteor showers are a popular time to try the meteor scatter mode.

Chapter 3

Finding Other Hams: Your Support Group

An Elmer, or mentor (see Chapter 5), is very useful in helping you over the rough spots every newcomer encounters. As your interests widen, though, you'll need additional help. Luckily, hundreds and hundreds of potential Elmers are in the myriad ham radio clubs and organizations around the world.

One of the oldest traditions of ham radio is helping the newcomer, and hams are great at providing a little guidance or assistance. You can make your first forays into ham radio operating much easier and more successful by taking advantage of those helping hands. This chapter shows you how to find them.

Radio Clubs

The easiest way to get in touch with other hams is through the local radio club. Radio clubs have been around as long as radio. The first clubs were just groups of like-minded experimenters who collaborated to build radios when the technology was raw and success by no means assured. Over time, the club grew in size and importance. Today, clubs range from small, focused groups to large clubs, with wide-ranging interests.

The following points hold true for most hams:

✔ **Most hams belong to at least one club, sometimes several:** Belonging to a general interest club as well as one or two specialty groups is popular.

✔ **Most local or regional clubs have in-person meetings:** Membership is drawn largely from a single area.

✔ **Specialty clubs are focused on activities:** Activities such as contesting, low-power operating, or amateur television, may have a much wider (even international) membership. See the "Specialty Organizations and Clubs" section, later in this chapter, to discover different specialty clubs.

✔ **Individual chapters may have no in-person meetings:** They may conduct meetings only over the air.

Clubs are great resources for assistance and mentorship. As you get started in ham radio, you'll find that you need a lot of basic questions answered. I recommend that you start by joining a general interest club (see the upcoming section). If you can find one that emphasizes assistance to the new ham, so much the better. You'll find the road to enjoying ham radio a lot smoother in the company of others.

Finding and choosing a club

To find ham radio clubs in your area, start at www.qrz.com/clubs.html and select your state to find a list of the state's radio club Web sites. The American Radio Relay League (ARRL) also has a directory of affiliated clubs on its Web site at www.arrl.org/FandES/field/club/clubsearch.phtml. Enter your state, city, or zip code to locate nearby clubs.

Focus on the general interest clubs and look for the clubs that offer help to new hams. For example, I found this listing for one of the largest Seattle, Washington, clubs through the ARRL Web site:

Name: MIKE & KEY AMATEUR RADIO CLUB

Specialties: General Interest

Call sign: K7LED

Services: Help for newcomers, entry-level classes, higher-level classes, other

This club is well suited for a new ham. You find yourself in the company of others recently licensed, so you won't feel self-conscious when asking questions. You have programs and activities to learn from and opportunities for you to contribute.

If you have more than one club available in your area, how do you make a choice? Consider these points when making a decision:

- ✔ **Which club has meetings that are more convenient for you?**

    Check out the meeting times and places for each club.

- ✔ **Which club includes activities or programs that include your interests?**

    If a club has a Web site or newsletter, review the past few month's activities and programs to see if they sound interesting.

- ✔ **Which club feels more comfortable to you?**

    Don't be afraid to attend a meeting or two to find out what the club is like.

You'll quickly find out that the problem is not finding clubs, but in choosing among the hundreds of them! Unless the club has a strong personal participation aspect, such as a public service club, you can join as many as you want just to find out about that part of ham radio. Most clubs have a newsletter and a Web site that give you a valuable window into one of ham radio's many specialties.

Participating in a club

After you pick a general interest club, show up for meetings, and make a few friends right away, how do you really start participating? Do ham radio clubs have a secret protocol? What if you goof up?

Obviously, you won't start your ham club career by running for president at your second meeting, but ham clubs are pretty much like every other hobby group. As such, you can become an insider with easy first steps. You're the new guy or gal, which means you have to show you want to belong. Here are some tips to help you assimilate:

- ✔ Show up early and help set up, make the coffee, hang the club banner, help figure out the projector, and so forth. Stay late and help clean up, too.

- ✔ Be sure to sign in, sign on, or sign up if you have an opportunity to do so, especially if it's your first meeting.

- ✔ Wear a name tag or other identification that announces your name and call sign in easy-to-read letters.

- ✔ Introduce yourself to whomever you sit next to.

- ✔ Introduce yourself to a club officer as a visitor or new member. If a "stand up and identify yourself" routine occurs at the beginning of the meeting, be sure to identify yourself as a new member or visitor. If other people also identify themselves as new, go introduce yourself to them later.

✔ After you've been to two or three meetings, you will probably recognize some of the club's committees or activities. If one of them sounds interesting, introduce yourself to whoever spoke about it and offer to help.

✔ Show up at club activities or work parties!

✔ Comb your hair. Brush your teeth. Sit up straight. Wear matching socks. Yes, Mom.

These magic tips are not just for ham radio clubs, but also for just about any club. And just like those other clubs, ham clubs have their own personality. They vary from wildly welcoming to tightly knit, seemingly impenetrable groups. Though after you break the ice, hams seem to bond for life. And when you're a club elder, remember to extend a hand to new members the way you appreciated when you were a new face yourself!

Getting involved

Okay, you're a regular! How can you get involved? In just about every ham club, you'll find the following jobs need doing. Find out who is currently in that position and offer your help. You'll discover a new aspect of ham radio, gain a friend, and make a contribution.

✔ **Field days:** Planners and organizers can always use a hand in getting ready for this June operating event. Offer to help with generators, tents, and food, and find out about the rest of it as you go. Helping out with field days is a great way to meet the most active members of the club.

✔ **Conventions or hamfests:** If the club hosts a regular event, almost any kind of help is always needed. If you have any kind of organizational or management expertise, so much the better.

✔ **Awards and club insignia:** Managing sales of club insignia is a great job for new members — keeping records, taking orders, and making sales at club meetings. If you have an artistic or crafts bent, don't be afraid to make suggestions.

✔ **Libraries and equipment:** Many clubs maintain a library of reference books or have equipment that is loaned to members. All you have to do is keep track of it and make it available to other members.

✔ **Club stations:** If your club is fortunate enough to have its own radio shack or repeater station, some maintenance work — such as working on antennas, changing batteries, tuning and testing radios, or just cleaning — always needs to be done. Buddy up with the station manager; you can become familiar with the equipment very quickly. You need not be technical, just willing.

If you can write or design Web sites, don't hesitate to volunteer your services to the club newsletter editor or Webmaster. Chances are that they have several projects backlogged and would be delighted to have your help.

Along with the ongoing club committees and business, you usually can find a number of club-sponsored activities. Some clubs are organized around one major activity while others seem to have one or two going on every month. Here are a few common club activities:

- ✔ **Public service:** This activity usually entails providing local communication during a sporting or civic event, such as a parade or festival. These events are great ways for you to hone your exchanging messages and operating skills.

- ✔ **Contests and challenges:** Operating events are great fun and many clubs enter on-the-air contests as a team or club. Sometimes, clubs challenge each other to see which can generate the most points. You can either get on the air yourself or join a multi-operator station.

- ✔ **Work parties:** What's a club for if not to help out its own members? Raising a tower or doing antenna work with other members is a great way to meet active hams and discover this important aspect of station building.

- ✔ **Construction projects:** Building your own equipment and antennas is a lot of fun, so clubs may occasionally sponsor group construction projects. Building your own equipment saves money and lets everyone work together to solve problems. If you like building things or have technical skills, here's a great way to help out.

Supporting your club by participating in activities and committees is important. For one thing, you can acknowledge the help you get from the other members. You also start to become a mentor (or *Elmer*) yourself to other new members. (See Chapter 5 for more on Elmers.) By being active within the club, you strengthen the organization, your friendships with others, and the hobby in general.

The ARRL

The *American Radio Relay League* (ARRL) is the oldest continuously functioning amateur radio organization in the world. Founded before World War I, the ARRL provides services to hams around the world and plays a key part in representing the ham radio cause to the public and governments. That ham radio could survive for nearly 100 years without a strong leadership organization is hard to imagine and the ARRL has filled that role.

I am devoting a separate section of this chapter to the ARRL simply because it is such a large presence within the hobby to U.S. hams (and with its sister organization, Radio Amateurs of Canada or RAC, to Canadian hams) and because of the many valuable services it provides, particularly to new hams.

ARRL's benefits to you

The most visible benefit of ARRL membership is that you receive *QST* magazine every month. *QST*, the largest and oldest ham radio magazine (shown in Figure 3-1), includes feature articles on both technical and operating topics; reports on regulatory information affecting the hobby; the results of ARRL-sponsored competitions; and numerous columns covering a wide variety of topics. All the largest ham radio equipment manufacturers advertise in *QST* as well. Other excellent ham radio magazines exist, but *QST* is the most widely read and important.

Figure 3-1: QST is the official magazine of the ARRL.

Along with the print magazine, the ARRL also maintains an active and substantial Web site at www.arrl.org. You can find the following features on the Web site:

- News and general interest stories posted every day
- The Technical Information Service, an extensive reference service, including technical document searches with numerous articles available online
- An active ham-radio swap and shop, open 24 hours a day
- A number of free e-mail bulletins and newsletters

The ARRL Field Organization coordinates the activities of hundreds of volunteers. While the League has a paid staff at its headquarters in Newington, Connecticut, the ARRL is by and large a volunteer organization. These volunteers are organized into 80 different sections in 15 divisions.

The Field Organization also includes the world's most extensive nongovernmental radio network, the *National Traffic System* (NTS). The NTS is key in rapidly responding to all sorts of emergencies and is active on a daily basis, passing radio messages, or *traffic,* all over the world.

Along with administrative and organizational functions, the ARRL is also the largest single sponsor of operating activities for hams. The ARRL sponsors numerous competitive events, called *contests,* award programs, and technical and emergency exercises.

ARRL's benefits to the hobby

By far the most visible aspect of the ARRL on the ham bands is the ARRL headquarters station, W1AW. Carrying the original call sign of ARRL founder Hiram Percy Maxim, the powerful station (shown in Figure 3-2) is a worldwide beacon, beaming bulletins and Morse code practice sessions to hams around the planet every day. Visiting hams can also operate the W1AW station themselves (don't forget your license!). Most hams think being at the controls of one of the most famous and storied ham stations in all the world is the thrill of a lifetime!

Figure 3-2:
The ARRL
station
W1AW in
Newington,
Connecticut.

The ARRL also provides the Volunteer Examiner Coordinator (VEC) service (you may have taken your licensing test during an ARRL-VEC session). With the largest number of volunteer examiners, the ARRL-VEC assists thousands of new and active hams take their licensing exams, obtain vanity and special call signs, renew their licenses, and update their license information of record. When the FCC could no longer maintain the staff to administer licensing programs, the ARRL and other ham organizations stepped forward to create the largely self-regulated VEC programs instrumental to today's healthy ham radio.

The least visible of its functions is the representation of the Amateur Radio service to governments and regulatory bodies — arguably one of the most important functions of the ARRL. Radio spectrum is prime territory in the telecommunications-driven world and many other services would like to get access to amateur frequencies, regardless of the long-term impact. The ARRL also serves as a trusted voice to the FCC and Congress, helping regulators and legislators understand the special nature and needs of the Amateur Service.

ARRL's benefits to the public

Although naturally focused on its membership, the ARRL takes seriously its mission to support Amateur Radio. The ARRL Web site is largely open to the public as are all bulletins broadcast by W1AW. The ARRL performs these services:

- ✔ **Facilitates the development of emergency communications:** In conjunction with the Field Organization, the Amateur Radio Emergency Service (ARES) teams around the country provide thousands upon thousands of hours of public service every year. While individual amateurs render valuable aid in times of emergency, the organization of these efforts multiplies its usefulness.

- ✔ **Publishes the ARRL Handbook:** Now in its 81st edition, the handbook is an acknowledged telecommunications industry reference, used by professionals as well as amateurs. The League publishes numerous technical references and guides, including conference proceedings and standards, furthering the state of the radio art.

- ✔ **Promotes not only ham radio, but also technical awareness, with strong support programs for elementary through secondary schools:** The League is also involved with the Boy Scouts and Girl Scouts Radio merit badge and Jamboree On-The-Air programs. For hams, the ARRL sponsors a growing series of Web-based continuing education courses.

Joining the ARRL

The ARRL is a volunteer-based, membership-oriented organization. Just like your local club, if the ARRL is to provide you with useful services and effectively represent your interests, you need to be a member. Better yet, the League needs you to be a volunteer! Rest assured, even as a new ham, you can make a meaningful contribution as a volunteer. To find out how to join, go to www.arrl.org/join.html.

Specialty Organizations and Clubs

Ham radio is big, wide, and deep. The hobby has many individual communities that fill the airwaves with diverse activities. A *specialty club or organization* focuses on one aspect of ham radio that emphasizes certain technologies or types of operation. Quite a number of specialty organizations have a worldwide membership, as well.

To find specialty clubs, search the Internet by entering your area of interest and the word **club**. For example, entering **10 meter club** at www.google.com turns up nearly a dozen ham clubs or forums.

Some clubs focus on particular operating interests, such as qualifying for awards or operating on a single band. An example of the latter is the 10-10 International Club (www.ten-ten.org/), which is for operators that prefer the 10 meter band, a favorite of low-power and mobile stations. The 10-10 Club sponsors several contests every year and offers a set of awards for contacting its members. The Six Meter International Radio Klub (SMIRK) promotes activity on the 6 meter band with its special and unusual methods of signal propagation. The club's contests and awards are listed at www.smirk.org/.

Another type of specialty club is the *contest club*. Members enjoy participating in competitive on-the-air events known as *radiosports,* or simply *contests* (the topic of Chapter 11). The clubs challenge each other, sponsor awards and plaques, and generally encourage members to build up stations and techniques to become top contest operators. Contest clubs tend to be local or regional due to the rules of club competition. One of the oldest U.S. clubs, the Yankee Clipper Contest Club, maintains a list of U.S. contest clubs at www.yccc.org/Links/Contest_clubs.htm. You can view an extensive list of clubs that compete in ARRL contests at www.arrl.org/contests/club-list.html, although many general interest clubs are included. For contest clubs around the world, try the list at www.ac6v.com/clubs.htm#DX.

No less competitive than contest operators are the long-distance communications specialists, or *DXers,* who specialize in long-distance contacts with places well off the beaten track. The quest to *work 'em all* (contact every country on every ham band) lasts a lifetime and leads you to learn obscure geography when a ham begins transmitting from a tiny island somewhere far away. DXers form clubs to share operating experiences and host traveling hams, fostering international communications and goodwill along the way. I belong to the Western Washington DX Club (www.wwdxc.org) with members worldwide, although most of them are near Seattle, Washington.

Because of an international nature, DX clubs tend to have members sprinkled around the globe. If one is in your area, you can find it at www.dailydx.com/clubs.htm.

Handi-Hams

Ham radio provides excellent communication opportunities to people who otherwise find themselves constrained by physical limitations. Handi-Hams, founded in 1967, is a specialty organization dedicated to providing tools that make ham radio accessible to hams with disabilities of all sorts. Handi-Hams helps those people turn disabilities into assets through ham radio.

Handi-Hams not only helps hams with disabilities reach out to the rest of the world, but also helps its members link up with other members and to various helpful services. Even if you're not disabled, Handi-Hams may be a welcome referral to someone you know or you may want to volunteer your services to Handi-Hams. The Handi-Hams Web portal provides links to an extensive set of resources at www.handihams.org.

AMSAT

AMSAT (short for Amateur Radio) is the international organization that helps coordinate satellite launches and oversees the construction of its own satellites. Radio operation through a satellite is a lot easier than you may think, as you can find out in Part IV.

Yes, Virginia, there really are Amateur Radio satellites up there whizzing through the heavens! The first one, launched in 1962, sent a Morse code beacon consisting of the letters "HI" (in Morse code speak, "di-di-di-dit, di-dit"), known as the telegrapher's laugh. Nowadays, more than 20 satellites are in orbit. The first, OSCAR-1 (Orbiting Satellite Carrying Amateur Radio), was about the size of a large coffee can. Modern satellites, such as the AO-40 shown in Figure 3-3, are full-sized and full-featured. The satellite pictured provides communications on several amateur UHF and microwave bands.

You can find the AMSAT Web site at www.amsat.org. The list of links at the bottom of the page is where you find information on the organization and membership.

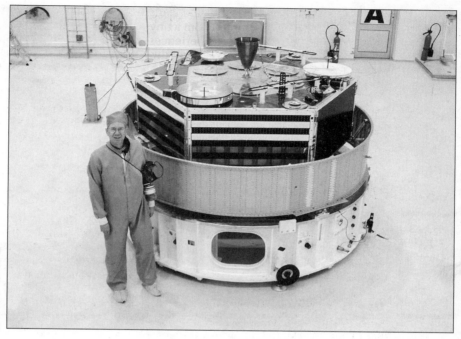

Figure 3-3:
Amateur
satellite
AO-40.

TAPR

Another acronym, TAPR stands for *Tucson Amateur Packet Radio* (www.tapr.org), which has been a force in digital communications development for more than 20 years. TAPR has been instrumental in bringing modern digital communications technology to ham radio. In return, TAPR members created several innovative communication technologies that are now commonplace beyond ham radio, such as the communications system known as *packet radio,* which is widely used in industry and public safety. More recently, TAPR members have been instrumental in implementing modern digital communications, such as *networking* and *spread-spectrum technology,* as shown in Figure 3-4. If you have a strong computer background, TAPR is likely to have activities that pique your interest.

Young Ladies' Radio League — the YLRL

The YLRL (Young Ladies' Radio League) was founded in 1939 and is dedicated to promoting ham radio to women, encouraging YL (young lady) activity on the air, promoting women's interests within the hobby, and providing an organization for female hams.

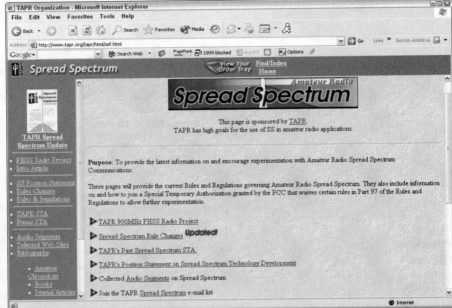

Figure 3-4:
TAPR is at the forefront of amateur digital technology.

YLRL is an international organization with chapters in many countries, hosting conventions and creating opportunities for travel. The YLRL's home page at www.ylrl.org provides a list of activities and member services. The YLRL has a vigorous award program and also sponsors on-the-air nets for members. It also sponsors on-the-air competitions for YLs and OMs (*Old Men,* or, in other words, male operators of any age).

QRP clubs: ARCI, AmQRP, and G-QRP

QRP is ham radio shorthand for *low power* operating, using just a few watts of power to span the oceans. Like bicyclists among the motorists, QRP enthusiasts emphasize skill and technique, preferring to communicate with a minimum amount of power. The largest U.S. QRP clubs are the QRP Amateur Radio Club International, known as QRP ARCI (www.qrparci.org/), and the American QRP Club, or AmQRP (www.a-qrp.org/). Both have excellent magazines, such as the *QRP Quarterly* seen in Figure 3-5, full of construction projects and operating tips.

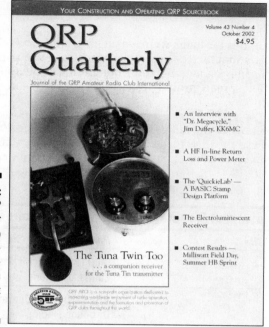

Figure 3-5: The QRP Amateur Radio Club International publishes this excellent quarterly.

Many QRP clubs are worldwide, and one of my favorites is the British club, G-QRP (*G* is the prefix of call signs in England). You find the G-QRP home page at www.gqrp.com/. If you like building your own gear and operating with a minimum of power, check out these clubs and other groups of QRPers.

Online Communities

Just like every other human activity, ham radio has online communities in which members discuss the various aspects of the hobby. They provide resources and support, available 24 hours a day. Will these communities replace ham radio? Not likely; the magic of radio is just too strong. By their presence, though, they make ham radio stronger by cementing relationships and adding structure.

Reflectors

The first online communities for hams were e-mail lists, known as reflectors. *Reflectors* are mailing lists that take e-mail from one mailbox and re-broadcast it to all members. With some list memberships numbering in the thousands, reflectors get information spread around pretty rapidly. Every ham radio interest has a reflector.

Table 3-1 presents a number of the larger Web sites that act as hosts for reflectors. You can browse the directories and decide which list suits your interests. Be careful you don't wind up spending all your time on the reflectors and none on the air!

Table 3-1	Hosts and Directories for Ham Radio Reflectors
Host Address or Web Site	*Partial List of Topics*
www.qth.com	Over 600 topics on radios, bands, operating, and awards
www.n4kss.net/reflectors.html	ARES, Portable Operation, Direction Finding
www.contesting.com	TowerTalk, CQ-Contest, Amps, Top Band (160-meters)
www.ipass.net/teara/reflect.html, www.columbia.edu/~fuat/cuarc/ mailing-lists.html, www.ac6v.com/ mail.htm	Directories of reflectors hosted on other sites

The new ham may want to join one of the Elmer e-mail lists that are set up specifically to answer questions and offer help. To find general and topical Elmer lists, enter **ham radio elmer reflector** into a Web search engine and you can turn up several candidates.

Because my main interests are QRP operating on the HF bands, contesting, and making long-distance or *DX* contacts, I subscribe to the QRP reflector, the CQ-Contest reflector, a couple of the DX reflectors, and the Top Band reflector about 160-meter operating techniques and antennas. To make things a little easier on my e-mail box, I subscribe in digest form, which means I get one or two bundles of e-mail every day instead of many messages. Most reflectors are lightly moderated and usually close the list to any posts not from subscribed members, or in other words, spam.

The reflectors at Yahoo! Groups (groups.yahoo.com/) offer a little more than just e-mail distribution. They also offer file storage, a photo display function, chat rooms, polls, and excellent member management. To take advantage of these services, you must create a personal Yahoo! profile first. More than 500 ham radio groups on Yahoo!Groups are in the Hobbies section. Some are open to the public and others require you to subscribe to the list. Log onto Yahoo! Groups and enter **ham radio** in the search engine to find them.

As soon as you settle into an on-the-air routine, subscribe to one or two reflectors. Reflectors are also a great way to find out about new equipment and techniques before you take the plunge and try them yourself.

Newsgroups

The USENET newsgroups encompass a stunning breadth of human interests, including ham radio. After you subscribe to a newsgroup, you can make any kind of post; the topics of discussions are wide-ranging.

You have the ability to subscribe to these newsgroups through most e-mail programs. Some services also manage newsgroup information, such as Google Groups (groups.google.com/). You can find ham radio topics in both the alt (eight topics) and rec (two topics) lists and have hundreds of messages every day. One popular example is the rec.ham-radio.swap group, which is a ham radio flea market that runs 24/7.

Newsgroups are generally unmoderated and open, so they tend to have a higher amount of spam and off-topic discussions than the moderated reflectors. Nevertheless, you can find a lot of excellent information in the newsgroups.

Portals

More than a reflector or meeting site, *portals* provide a comprehensive set of services and function as a ham radio home page. They feature news, informative articles, radio buy-and-sell pages, links to databases, reflectors, and many other useful services to hams. The best-known portals are eHam.net (www.eham.net) and QRZ.com (www.qrz.com).

QRZ (the ham radio abbreviation for "Who is calling me?") evolved from a call sign lookup service — what used to be a printed book known as a *callbook* — to the comprehensive site that you see today. The callbook features are incredibly useful and a number of call sign management functions are on the site. Ham radio news is the main forum on the home page, but Q&A forums and equipment exchanges are also on the portal.

The eHam.net site evolved from a specialty Web site known as Contesting.com, which focused on ham radio competitions (now encompassed within the eHam.net site). The eHam home page is organized around community functions, operating functions, and resources. You can find real-time links to a DX-station spotting system (frequencies of distant stations currently on the air) and the latest solar and ionospheric data that affect radio propagation.

I recommend that you bookmark both of these sites because they have a lively collection of news and articles along with useful forums and features. Both are gateways to e-mail reflectors that make subscription easy.

Hamfests and Conventions

Depending on how much you like collecting and bargaining, I may have saved the best for last. Ah, the hamfest! Hamfests are one of the most interesting events in ham radio. Imagine an Oriental bazaar crammed with technological artifacts spanning nearly a century, old with new, small with massive, tubes, transistors, computers, antennas, batteries . . . I'm worn out just thinking about it! I love a good hamfest; can you tell?

Hamfests are ham radio flea markets. The primary sellers are individuals vending treasures from a folding table or a tailgate. You may even find some commercial vendors. Some hamfests are very small and last only a few hours, while a few attract thousands over several days. Hamfests may be held indoors or outdoors.

A ham radio convention has a much broader slate of activities than a hamfest; it may include seminars, speakers, licensing test sessions, and commercial vendors. Conventions often include a swap meet along with the rest of the functions. Conventions often have a theme, such as emergency operations, QRP or low-power operating, or digital radio transmissions, to name a few.

Finding hamfests

In the United States, the best place you can find hamfests at is the ARRL Web site (www.arrl.org). At the top of the home page is a <u>Hamfests</u> link, which takes you to the Hamfest and Conventions page. Search for events by state, ARRL section, or ARRL division. Over 100 hamfests are usually listed at all times. Another good source of information is your ARRL Section News (look for the <u>ARRL Sections</u> link at the top right of the ARRL home page) or club newsletter. Most populated areas have several good-sized hamfests every year, even in the dead of winter.

After you have a hamfest in your sights, set your alarm for early Saturday morning (most are Saturday-only events) and get ready to be there at the opening bell. Be sure to bring the following things:

- ✔ **An admission ticket:** You need a ticket, sold at the gate or by advance order through a Web site or e-mail.

- ✔ **Money:** Take cash, because most vendors do not take checks or credit cards.

- ✔ **Something to carry your purchases in:** Take along a sturdy cloth sack, backpack, or other bag that can tolerate somewhat grimy, dusty electronic devices.

- ✔ **A hand-held or mobile rig:** Most hamfests have a talk-in frequency, which is almost always a VHF or UHF repeater. If you're unfamiliar with the area, you can get directions while en route.

  If you go with a friend and both of you take hand-held radios, you can share tips about the stuff you find while walking the aisles.

- ✔ **Water and food:** Don't count on food being available, but the larger hamfests almost always have a hamburger stand. Rarely is gourmet food on hand, but expect the same level of quality as that of the concession stands at a ballgame. Taking along a full water bottle is a good idea.

Buying at hamfests

After parking, waiting, and shuffling along in line, you finally make it inside the gates. You're ready to bargain! All hamfests are different, but here are some guidelines to live by, particularly as a novice hamfest customer:

- ✔ If you're new to ham radio, buddy up with a more experienced ham to steer you around hamfest pitfalls.

- ✔ Most prices are negotiable; more so after lunch, but a good deal goes quickly. Most vendors are not interested in trades, but you do no harm by offering.

- ✔ Hamfests are good places to buy accessories for your radio, often selling for a fraction of the manufacturer's price if separate from the radio. Commercial vendors of new batteries often have good deals on spare battery packs.

- ✔ Many hamfests now have electricity available so that vendors can demonstrate equipment. If a vendor refuses to demonstrate a supposedly functional piece of gear, or won't open up a piece of equipment for inspection, you may want to move along.

✔ Unless you really know what you are doing, avoid antique radios. They often have quirks that can make using them a pain or require impossible-to-get repair parts.

✔ Be familiar with the smell of burnt or overheated electronics, especially transformers and sealed components. Direct replacements may be difficult to obtain.

✔ Don't be afraid to ask what something is. Most of the time, the ham behind the table enjoys telling you and, even if you don't buy it, the discussion may attract a buyer.

✔ If you know exactly what you are looking for, check the auction Web sites and radio swap sites, such as www.eham.net, www.qrz.com, and www.arrl.org/RadiosOnline/, before you attend the hamfest. You can get an idea of the going price and average condition, so you're less likely to get gouged.

Other activities at hamfests

Along with buying and selling, many hamfests also have programs and speakers and even license exam sessions, like small conventions. Look for a flyer or check the hamfest Web site for information about special services that may be available.

Finding conventions

You can find out about upcoming conventions in the same places as hamfests. Conventions tend to be more extravagant affairs than hamfests and may be advertised in ham radio magazines as well as online. At a convention, the main focus is on programs, speakers, and socializing. Conventions are usually held in a hotel.

The two largest ham radio conventions of all are the Hamvention, held in Dayton, Ohio, in mid-May, and the International Amateurfunk-Ausstellung, held in Friedrichshafen, Germany, in late June. Dayton regularly draws 25,000 or more and Friedrichshafen nearly that many. Both have mammoth flea markets, an astounding array of programs, internationally-known speakers, and you can't possibly see all of it. If at all possible, you should go to one of them at least once in your life!

The ARRL National and Division Conventions (listed on the ARRL Web site) are held in every region of the United States. The Radio Amateurs of Canada also host a national convention every year. These conventions typically attract a few hundred to a few thousand and are designed to be family friendly. They also provide a venue for specialty groups to host conferences within the overall event. These smaller conferences are where you find extensive programs on direction finding, QRP, county hunting, wireless networking on ham bands, and so on.

Some conventions and conferences emphasize one of ham radio's many facets, such as DX-ing, VHF and UHF operating, or digital technology. If you are a fan of a certain mode or activity, treating yourself to a weekend convention is a great way to meet some of the other hams that share your tastes and discover more about your interests. Table 3-2 lists a few of the specialty conventions held around the United States each year.

Table 3-2	Specialty Conventions	
Name	*Theme*	*Web Site*
Visalia International DX Convention (hosted alternately by the Northern and Southern California Contest Clubs)	DX & Contesting	www.scdxc.org/ visalia/index.html
W9DXCC (hosted by the Northern Illinois DX Association)	DX & Contesting	www.w9dxcc.com
Pacific Northwest DX Convention (rotates between Seattle, WA; Portland, OR; and Vancouver, BC, each year)	DX & Contesting	www.wwdxc.org
Southeastern VHF Conference (hosted by the Southeastern VHF Society)	VHF, UHF, and Microwaves	www.svhfs.org
International EME Conference	EME (Earth-Moon-Earth) operating	www.qsl.net/ eme2004
Microwave Update (sponsored by the Pacific NW VHF Society)	VHF, UHF, and Microwaves	www.microwave update.org/
Digital Communications Conference (hosted by the ARRL and TAPR)	Digital Communications	www.tapr.org/ conferences.html

Part II

Wading through the Licensing Process

The 5th Wave   By Rich Tennant

"Okay — now that the paramedic is here with the defibrillator and smelling salts, prepare to open your test booklets..."

In this part . . .

Getting a license from (drum roll, please) the Federal Government (trumpet flourish) sounds daunting, but getting a license has never been easier. In fact, you don't even have to deal with a government agency until your license shows up in the mail! Ham volunteers do all the licensing work — teaching, testing, and filing paperwork. If you can fill out the application for a driver's license, you can get a ham radio license.

In this part, you find out all about how hams get their licenses. Odds are you'll need an Elmer to guide you along the way, of course, and I introduce you to Elmer here. I explain just how the test is conducted, how to study for it, and how to sign up for a test session. After you pass (naturally!) I spill the beans about how to find out what your new call sign is. You can even pick out your own call sign if you want!

Chapter 4

Figuring Out the Licensing System

*U*nlike some of the other types of radios available to the public, ham radios can *only* be used with a license. But how do you get a license? How much does it cost? Can you obtain different kinds of licenses? Like most people, you're probably familiar with the process of getting a license to drive your car, to fish, or to get married. Ham radio licensing is a little different, but easy to deal with once you know how it works.

Amateur Radio is one of many different types of radio services that use the radio waves to communicate. Other services include Broadcast (AM, FM, TV), Public Safety (police and fire departments), Aviation, and even Radar. To maintain order on the airwaves, the Federal Communications Commission (FCC) requires that each station be licensed. Stations in all the different services must abide by the regulations created by the FCC in order to obtain and keep their licenses, which give them permission to transmit according to the rules for that service. That's what a ham license is — authority for you to transmit on the frequencies that licensees of the Amateur Radio Service are permitted to use. This chapter explains the FCC's licensing system for Amateur Radio in the United States.

The Amateur Service: An Overview

By international treaty, the Amateur Service in every country is a licensed service. That is, a government agency has to approve the application of every ham to transmit. Although regulation seems a little quaint given all of the communications gadgets for sale these days, licensing is necessary for a couple of reasons:

- ✔ It allows amateurs to communicate internationally and directly, without using any kind of intermediate system that regulates their activities.
- ✔ Because of the power and scope of Amateur Radio, hams need a minimum amount of technical and regulatory background so they can co-exist with other radio services, such as broadcasting.

By maintaining the quality of licensees, licensing helps ensure that the Amateur Service makes the best use of its unique citizen access to the airwaves. Amateur Service sets ham radio apart from the unlicensed services and is recognized in the FCC rules as the Basis and Purpose of the Amateur Radio Service, rule 97.1:

- ✔ Recognition of ham radio's exceptional capability to provide emergency communications (rule 97.1(a))
- ✔ Promote the amateur's proven ability to advance the state of the radio art (rule 97.1(b))
- ✔ Encourage amateurs to improve their technical and communications skills (rule 97.1(c))
- ✔ Expand the number of trained operators, technicians, and electronics experts (rule 97.1(d))
- ✔ Promote the amateur's unique ability to enhance international goodwill (rule 97.1(e))

Pretty heady stuff, eh? Ham radio does all these good things in exchange for radio space among the commercial and government users. You can find all the rules for Amateur Radio at wireless.fcc.gov/rules.html (click the Part 97 link for the Amateur Radio rules). Plain English discussion of the rules is available on the ARRL Web site at www.remote.arrl.org/FandES/field/ regulations/rules-regs.html or in the ARRL's *FCC Rule Book*.

Frequency allocations

To keep order in the growing radio communications field, countries got together in 1932 and formed the International Telecommunication Union (www.itu.int/home/), based in Geneva, Switzerland. The ITU is not part of any government, but is a forum for deciding and recording the rules of radio spectrum usage. The ITU divided the spectrum into small ranges in which specific types of uses occur (see Figure 4-1). These ranges are called *frequency allocations.*

The world is divided into three regions, as follows:

- ✔ **Region 1:** Europe, Russia, and Africa
- ✔ **Region 2:** North and South America
- ✔ **Region 3:** Asia, Australia, and most of the Pacific

Within each region, each type of radio service — amateur, military, commercial, and government — is allocated a share of the available frequencies. Luckily for amateurs, most of their allocations are the same in all three regions. Figure 4-2 shows the HF range frequencies (from 3 to 30 MHz). This allocation is very important, particularly on the long-distance bands where radio signals might propagate all the way around Earth. Talking to someone in a foreign country is pretty difficult if you don't use the same frequency!

To get an idea of the complexity of the allocations, browse to the Region 2 allocation chart at www.ntia.doc.gov/osmhome/allochrt.html. (If you have Adobe Acrobat you can download the entire color chart.) The individual colors represent different types of radio services. Each service has a small slice of the spectrum, including amateurs. Can you find the Amateur Service on the chart? *Hint:* It's green.

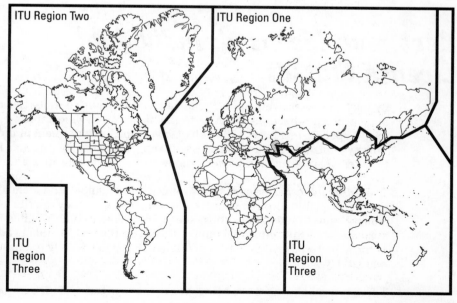

Figure 4-1:
ITU region
map
showing the
world's
three
telecom-
munications
administra-
tive regions.

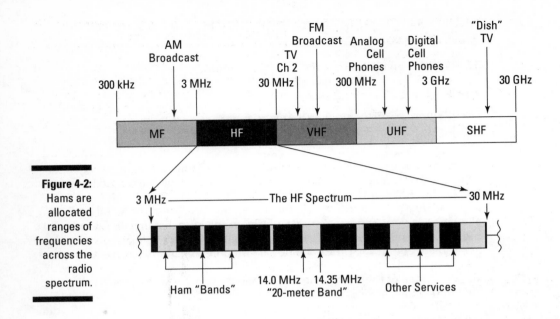

Figure 4-2:
Hams are allocated ranges of frequencies across the radio spectrum.

Amateurs have small allocations at numerous places in the radio spectrum and access to those frequencies depends on the type of license class you possess. The higher class license that you have, the more frequencies you can use!

Becoming Licensed: Individual License Classes

By taking progressively more challenging exams, more frequencies and operating privileges become available to the ham, as shown in Table 4-1. After you pass a specific test level, called an *element,* you have permanent credit for it. This system allows you to progress at your own pace. Your license is good for a ten-year period and you can renew it without taking an exam.

The ARRL and other organizations publish study guides and manuals. Some may be available through your local library. (Be sure they are the latest version, because the test questions change from time to time.) Online tests are available with the actual questions that are on the test. By taking advantage of these materials, you'll have confidence that you're ready to pass the exam on test day.

Three types of licenses are being granted today: Technician, General, and Amateur Extra.

Table 4-1	Privileges by License Class	
License Class	*Privileges*	*Notes*
Technician	All amateur privileges above 50 MHz	Passing 5 wpm Morse code test adds the privileges of the grandfathered Novice class.
General	Technician privileges plus most amateur HF privileges	
Amateur Extra	All amateur privileges	Small sub-bands are added on 80, 40, 20, and 15 meters.

The Amateur Radio service is undergoing some restructuring as I'm writing this book. Some requirements, privileges, and even license class names may change. Be sure that you have the current version of study materials that reflect the correct rules and regulations.

Technician class

Nearly every ham starts with a Technician-class license, also known as a *Tech* license. The Technician licensee is allowed access to all 17 ham bands with frequencies of 50 MHz or higher. These privileges include operation at the maximum legal limit and all types of communications. Tech licensees, on the other hand, may not transmit on the bands below 30 MHz.

The test for this license consists of 35 multiple-choice questions on regulations and technical radio topics. You have to get 26 or more correct to pass.

If you pass a 5-word-per-minute (wpm) Morse code exam, you also receive some transmitting privileges in segments of four of the traditional shortwave HF bands, as shown in Table 4-2. You can take the Morse code exam at the same test session as the written exam or at a later date if you like.

Table 4-2	Additional Privileges with Morse Code Exam
Band	*Frequency Privileges*
80 meters	3.675–3.725 MHz (Morse code only)
40 meters	7.100–7.150 MHz (Morse code only)
15 meters	21.100–21.200 MHz (Morse code only)
10 meters	28.100–21.300 MHz (Morse code, RTTY, and data); 28.300–28.500 (Voice)

Morse code is required for amateur operation below 30 MHz because of an international treaty adopted many years ago. At that time, a great deal of commercial and military radio traffic — news, telegrams, ship-to-ship, and ship-to-shore messages — was conducted using the code. Back then, using Morse code was considered a standard radio skill. This part of the treaty was recently dropped, and Morse code may yet lose its place in amateur licensing. Nevertheless, Morse code still makes up a great deal of amateur operations — from casual ragchewing, to passing messages, contests, and emergency operations. Its efficient use of transmitted power and spectrum space, as well as its innate musicality and rhythm, make it very popular with hams.

General class

By passing the 5-wpm Morse code exam and another 35-question written exam, the doors of Amateur Radio are flung wide open. General class licensees have full privileges on nearly all amateur frequencies with only small portions of some HF bands off limits. The General class exam covers many of the same topics as the Technician exam, but in more detail. Some new topics a more experienced ham is expected to understand are on the General class exam.

After starting with the entry-level Technician license, most hams today upgrade to the General class. When you obtain a General class license, you've reached a great milestone. All of the important frequencies on the HF bands are available to General class licensees.

Amateur Extra class

General class licensees still can't access everything; the lower segments of several HF bands are for Extra class licensees only. These segments are where the expert operators hang out. These segments are prime operating territory. If you become interested in contesting, contacting rare foreign stations *(DX-ing),* or just having access to these choice frequencies, you want to get your Amateur Extra license, the top level of all license classes.

The Amateur Extra exam consists of 50 multiple-choice questions, 37 of which you must answer correctly to pass. The exam covers additional rules and regulations associated with sophisticated operating and several advanced technical topics. Hams that pass the Amateur Extra exam consider it a real feather in their caps. Do you think you can climb to the top rung of the licensing ladder?

Grandfathered classes

The Amateur Radio Service licensing rules have changed in the past 15 years to reduce the number of license classes. Hams that hold those licenses in deleted classes may renew those licenses indefinitely, but no new licenses for those classes are being issued. Three "grandfathered" license classes exist:

- ✔ **Novice:** The Novice license was introduced in 1951 with a simple 20-question test and 5-wpm code exam. A General class (or higher ham) administered the exam. Originally, the license was good only for a single year, at which point the Novice upgraded or left the air. These days, the Novice license, like other licenses today, has a ten-year term and is renewable. Novices are restricted to segments of the 3.5, 7, 21, and 28 MHz amateur bands.

- ✔ **Technician-Plus:** These licensees are just the same as the current Technician licenses that passed the 5-wpm code test. The only difference is that a separate formal license class no longer exists.

- ✔ **Advanced Class:** Advanced Class licensees passed an exam midway in difficulty between those for the General and Amateur Extra classes and received frequency privileges beyond those for a General, but not as extensive as an Amateur Extra.

Table 4-3 shows the relative populations of each type of license holder.

Table 4-3	Relative Populations of Each License Class	
License Class	*Number of Active Licenses*	*Percent of Total Licensees*
Technician	277,063	37%
Technician-Plus (grandfathered)	72,467	10%
Novice (grandfathered)	41,114	6%
General	145,275	20%
Advanced (grandfathered)	85,420	12%
Amateur Extra	106,251	15%
Total	727,590	100%

*(**Source:** FCC database, 14 September 2003)*

Understanding Call Signs

Along with your license comes a very special thing — your call sign (or *call* to hams), which becomes your on-the-air identity. Each license granted by the FCC receives a unique call sign. Most hams change call signs once or twice before settling on one. Sometimes, your call sign starts taking over your off-the-air identity, too, as you become "Ward, NØAX" with your call sign in place of a last name. I have ham friends for whom I really have to think hard to remember their last name! Each letter and number in a call sign is pronounced individually and not as a word, for example "N Zero A X," not "No-axe."

Hams use the Ø symbol to represent the number zero, a tradition from commercial operating practices.

Ham radio call signs around the world are constructed from two parts: the prefix and the suffix. The suffix of a call sign, when added to the prefix, positively identifies you. Each call sign is unique. A lot of call signs consist of NØ and AX, but only one call sign is NØAX.

The *prefix* is composed of one or two letters and one numeral. For instance, the prefix in my call sign is NØ. It also identifies the country that issues your license and may also specify where you live within that country. For U.S. call signs, the numeral indicates the *call district* of your license when it was issued. Mine was issued in St. Louis, Missouri, which is part of the tenth or zero district. *Suffixes* consist of one to three letters. (No punctuation characters are allowed, just A to Z and Ø to 9.) The suffix in my call sign is AX.

The ITU assigns each country a block of prefix character groups that allows the government to assign licenses in all of its radio services. U.S. licensees (not just hams) all have call signs that begin with the letters A, K, N, or W. Even broadcast stations have call signs such as KGO or WLS. Most Canadian call signs begin with VE. English call signs may begin with G or M. Germans use D (for Deutschland), call signs that begin with J are Japanese, and so on. You can find the complete list of ham radio prefix assignments at www.ac6v. com/prefixes.htm#PRI.

Whatever class license you have is reflected in your call sign. When you get your first license, the FCC assigns you the next call sign in the heap for your license class — in much the same way you're assigned a license plate at the DMV. But, also like license plates, you can request special *vanity call signs,* within the call sign rules, of course. The higher your license class, the shorter and more distinctive your chosen call sign can be! Turn to Chapter 7 for more information on how to find out your call sign.

Your call sign is both a certification that you have personally passed the licensing exam and permission to construct and operate a station — a very special privilege.

The Volunteer Licensing System

While the processing of commercial and military license applications is handled directly by the FCC, it no longer administers Amateur Radio licensing examinations. In the United States, ham radio license exams are given by volunteer examiners who are certified by a coordinating organization: the VEC. This system is typical of the ham tradition of self-policing and self-organizing. The quality of the examination process is quite high. The flexibility provided by an all-volunteer system makes taking the test easy for the test-taker as well as the test-giver. The volunteers handle all of the paperwork and even file the results with the FCC. The actual license is still granted by the FCC, however.

In the Olde Days, tests were taken at the "local" FCC office, which could be hundreds of miles away. I vividly remember making long drives to the government office building to take my exams with dozens of other hams. Nowadays, the tests are usually available a short drive away at a club, school, or even in private homes. As a volunteer examiner, I've given over 30 exams around my kitchen table for hams as young as 10 years old!

Volunteer Examiner Coordinator (VEC)

The *Volunteer Examiner Coordinator* (VEC) is the organization that takes responsibility for coordinating the volunteer examiners who run the exam sessions and it also processes all of the FCC-required paperwork. The VEC with the most volunteer examiners is the nationwide group run by the American Radio Relay League (ARRL-VEC), but 13 other VECs are located around the United States. You can find a VEC near you at `wireless.fcc.gov/services/amateur/licensing/vecs.html`. Some VECs, such as the ARRL-VEC and W5YI-VEC, operate nationwide, while others work in only a single region.

The VEC is responsible for preparing and administering the license exams and other materials. It collects the test results, resolves all discrepancies, and then files all of the data with the FCC electronically. This process gets you your license and call sign much faster than if handled by the FCC. New licensees know their call signs within seven to ten business days from the time they take the test!

The VECs maintain a list of volunteer examiners (VEs), upcoming test sessions, and other resources for the ham test-taker. The VECs also deal with more than just exams — they can help you renew your license and change your address or name.

Volunteer Examiners (VEs)

The volunteer examiners make the system run. Each exam requires that three VEs are present and all three sign off on the test paperwork. VEs are responsible for all aspects of the testing process, including the meeting space and announcing the test sessions.

Any licensed amateur can become a VE by contacting one of the VEC organizations and completing whatever qualification process the VEC requires. In the case of the ARRL-VEC, a booklet on the volunteer licensing system is provided and the VE applicant must pass a short exam.

VEs are authorized to administer or *proctor* license exams at the same or lower class of license they hold themselves. For example, a General class VE can administer Technician and General exams, but not Amateur Extra.

How much does the exam cost? Under the ARRL-VEC, the current cost of attending a test session and taking an exam for any of the licensing elements is $12. If the volunteers incur any expenses, such as for supplies or renting the facility for testing, they're allowed to keep up to $6 per person (some choose to keep none of the fee). The remainder goes to the VEC to cover its expenses.

VEs are amateurs just like you (or just like you will be) and do a real service to the amateur community by making the licensing system run smoothly and efficiently. Don't forget to say, "Thanks!" at the conclusion of your test session, pass or fail. Better yet, become a VE yourself!

Chapter 5

Studying for Your License

You've decided to take the plunge and get your ham radio license! Congratulations! While you can't just run down to the store, buy your gear, and fire it up, becoming licensed is not a terribly difficult process. You can use a lot of resources as you prepare yourself for the ham radio exam. This chapter gives you some pointers on how best to make those preparations so that you enjoy studying and do well at test time.

If you buddy up with a study partner, studying is much easier. Having a partner helps you both stick with it! You each find different things to be easy and difficult, so you can work together to get through the sticky spots. Best of all, you can celebrate passing the test together!

Demystifying the Test

In order to do the best job of studying, you need to understand just what the test consists of and how it's designed. The tests for all license classes are multiple choice. You won't find any essay or pictorial questions. No oral questions of any kind are on the test either — no one asks you to recite the standard phonetic alphabet or sing a song about Ohm's Law.

The test for each license class (except Morse code) is called an *element*. The 5-words-per-minute Morse code exam, Element 1, tests your ability to receive Morse code. The written exam for the Technician license is Element 2. The written exams for General and Amateur Extra licenses are Elements 3 and 4, respectively.

During your studies, you'll encounter questions from the question pool, which is a large set of actual questions used on the test, all available to the public. The exam that you take is made up of a random selection of questions from that pool.

Well, the selection isn't quite random. The test covers four basic areas:

- ✓ **Rules & Regulations:** Important rules of the road that you have to know to operate legally
- ✓ **Operating:** Basic procedures and conventions hams follow on the air
- ✓ **Basic Electronics:** Elementary concepts about radio waves and electronics with some very basic math involved
- ✓ **RF Safety:** Questions about how to operate and install transmitters and antennas safely

The exam has a certain number of questions from each area; these questions are randomly selected from these areas. The Technician and General tests have 35 questions, while the Amateur Extra test has 50. If you answer three-quarters of the questions correctly, you pass!

Because the exam questions are public, you'll experience a strong temptation to just memorize the questions. Don't! Take the time to understand as much of the material as you can. When you do get your license, you'll find that studying pays off.

To be sure, memorization is required for a couple of areas, such as the lists of permitted frequencies for each license class. You have to know the exact limits of your license privileges before you can start operating.

Finding Resources for Study

If you're ready to start studying, what do you study? Are there books to study? Videos? Help! Lucky for you, the aspiring ham, numerous study references fit every taste and capability. Study aids commonly include classes, books, software, video, and online help.

Before purchasing any study materials, be aware that the test questions and regulations change infrequently, but they do change. The latest change was to the Technician class questions on July 1, 2003. Be sure that any study materials you purchase include the latest updates. The ARRL Web site shows the dates of the current question pools at www.remote.arrl.org/arrlvec/pools.html.

Finding licensing classes

If you learn better with a group of other students, you'll find classes beneficial. You can find classes by:

- ✔ **Asking at your radio club.** You can take classes sponsored by the club. If you don't see the class you want, contact the club by e-mail and ask about classes. If you need to find a club in your area, turn to Chapter 3.

- ✔ **Looking for upcoming exams to be held in your area.** The ARRL Web site has a search engine devoted to upcoming exams at `www.arrl.org/arrlvec/examsearch.phtml`, as does the W5YI test coordinator Web site (`www.w5yi.org`).

  Get in touch with the exam's contact liaison and ask about licensing tests. Because tests are often given at the end of class sessions, these contacts are frequently class instructors themselves!

- ✔ **Asking at a ham radio or electronics store.** If a ham radio store is in your vicinity (look in the Yellow Pages under "Electronic Equipment & Supplies" or "Radio Communication Equipment & Systems"), usually an associated bulletin board or Web site lists upcoming classes.

  Businesses that sell electronic supplies to individuals, such as Radio-Shack, may also know of classes. In a pinch, you can do a Web search for **ham radio class** or **radio licensing class** (or close variations) plus your town or region.

Other options for finding classes include local disaster-preparedness organizations, schools and colleges, which often provide space for classes, and public safety agencies such as the police and fire departments. By asking around you can usually turn up a reference to someone involved with ham radio licensing.

Occasionally you see classes advertised that take you from interested party to successful exam taker in a single weekend! The Technician exam is simple enough that a focused, concerted effort over a couple of days can cram enough material into your brain for you to pass the test.

The good part about these sessions is that by committing a single weekend, you can walk out the door on Sunday night having passed your exam and find your new call sign in the FCC's database the following week. For busy folks or those in a hurry, this time savings is a tremendous incentive.

Remember when you crammed for a final exam overnight and the minute after you took the exam you forgot everything that was on it? The same phenomenon applies to the licensing test as well. A lot of information you memorize in such a short amount of time can fade quickly over time. In two days, you can't really absorb the material well enough to permanently understand it. Unlike high school geometry, you use everything on the test later in real life. That means you have to *really* learn it, if not now, then again later. If you have the time to take a weekly course, that's the better choice.

Books, software, and videos

Of the available licensing study guide books, the best known guide is the ARRL's *Now You're Talking!* Aimed at the person studying for a Technician exam, it goes well beyond presenting just the questions from the question pool; *Now You're Talking!* attempts to really teach the why and how of the material. The ARRL books are available via www.arrl.org and from numerous retail outlets.

Gordon West WB6NOA has also written a series of licensing guides for all three license classes. These focus tightly on the pool in a question-and-answer format and are geared to the student who wants to pass the test quickly, so a lot of the background present in the ARRL's book is omitted. Gordon's books are sold through various retail outlets such as RadioShack.

In video format, the most detailed packages are the ARRL's *Technician Class* and *General Class* series of tapes. Each package is professionally prepared and has three tapes with a study guide. The companion video to the license study guide by Gordon West is the "No Code Video Seminar." CQ Magazine (www.cq-amateur-radio.com) offers an inexpensive series of *Getting Started In* videos that cover various aspects of ham radio. *Basic Technology for the Amateur Radio Enthusiast* by Alpha Delta Communications is another introductory video seminar that helps explain some of the technology involved with radio.

In software, one of the better series is the *Ham University* CDs, which includes a Technician licensing course and a Morse code disc. Several of the available courses have been converted from books to software packages by the same name.

Figure 5-1 shows some of these resources.

Online

Online resources are numerous, although generally not as thorough as the book and video courses. Nevertheless, the online practice exams can be particularly useful. When tutoring students, I urge them to practice the online exams repeatedly. Because the online exams use the actual questions, it's almost like the real thing. Practicing with them reduces your nervousness and gets you used to the actual format.

Figure 5-1:
A few of the numerous license guides and tutorial courses that are available as books, videos, and CDs.

The sites score your exams and let you know which of the study areas need more work. When you can pass the online exams by a comfortable margin every time, you'll do well in the actual session. You can find online exams at

✔ www.aa9pw.com/radio/

✔ www.qrz.com

✔ www.eham.net

While you're using the online tests, avoid the temptation to just take them over and over until you memorize them. Use the online tests as a way to test your understanding and to become familiar with the test format.

Finding a Mentor

Studying for your license may take you on a journey into unfamiliar territory. You can easily get stuck at some point — either on a technical concept or maybe a confusing regulation.

My Elmer experience

When I started in ham radio, my Elmer was Bill WNØDYV (now KJ7PC), a fellow high-school student who had been licensed for a year or so. I wasn't having any trouble with the electronics, but I sure needed a hand with the Morse code and frequency listings. I spent every Thursday over at Bill's house practicing Morse code *(pounding brass)* and learning to recognize my personal nemesis characters: D, U, G, and W. Without Bill's help, my path to getting licensed would have been considerably longer. Thanks, Bill! Since getting my license, I've required the assistance of several other Elmers as I entered new aspects of ham radio. If you can count on the help of an Elmer, the road to a license is much smoother.

As in many similar situations, the best way to solve a problem is to call on a more experienced person who can help you over the rough spots. In ham radio, these friendly mentors are called *Elmers*.

Rick Lindquist N1RL traces the history of the term *Elmer* — meaning someone who provides personal guidance and assistance to would-be hams — to *QST* magazine in a March 1971 "How's DX" column by Rod Newkirk W9BRD. Elmer didn't refer to anyone specific; just the friendly, more experienced ham who was around to help someone get their license and then get on the air. Nearly every ham has an Elmer at some point.

You won't get far by putting out a personal ad, but a lot of potential Elmers are out in Ham Radio Land. You can find Elmers in the following places:

- ✔ **A ham radio licensing class:** Often sponsored by a local ham radio club, a class is well worth whatever nominal fee, if any, charged just for the personal instruction you get and the ability to ask questions.

- ✔ **Radio clubs:** Radio clubs can help you find classes or may even host them. Clubs welcome visitors and often have an introduction session during the meeting. This session gives visitors an opportunity to say, "Hi, my name is so-and-so. I don't have a license yet, but I'm studying and might need some help." Chances are, you'll get several offers of assistance and referrals to local experts or classes. (Find out more about clubs in Chapter 3.)

- ✔ **Online:** You can ask questions and help others in their studies on numerous online sites. Teen Hams at www.youthtech.com/hamradio/ is popular. You can also find a good general-focus chat room at www.qsl.net/n5sdd/.

✓ **In your community:** Many of today's hams find their Elmers by looking around their own neighborhoods. Maybe a ham with a tower and antenna lives near you, or you see a car with a ham radio license plate at work. If you get the opportunity, introduce yourself and explain that you're studying for a license. Chances are that the person you're talking to needed an Elmer himself, way back when, and can give you a hand or help you find one.

After you get your license, you're in an excellent position to help other newcomers because you know just exactly how they felt at the start of the journey. Even if you're just one step ahead of the person asking the questions, you can be an Elmer. Some hams enjoy Elmer-ing so much that they devote much of their ham radio time to the job. You won't find a higher compliment in ham radio than "My Elmer."

Mastering Morse Code

Mastering Morse code is a very personal thing, such as playing an instrument or achieving a new athletic maneuver. Many people liken it to studying another language, with the same sudden breakthroughs separated by periods of repetition. Becoming a skilled Morse code operator results in a great sense of accomplishment, and you'll never regret it.

If you decide to study Morse code, some methods are much better than others. Avoid any method that encourages you to think of a table of character patterns. Managing the required 5 words per minute while looking up each character in your mind is difficult. After you can receive code as fast as these methods permit, you'll find moving to the higher speeds that make Morse code fun hard.

The style that most hams are successful with is the *Farnsworth Method*. The dits and dahs of each character are sent at the *code speed* (words per minute) you want to achieve, but the individual characters are spaced far enough apart in time *(character spacing)* so that the overall word speed is low enough for you to process the character's sound pattern. For the beginner, a common sending speed for individual characters is 7 words per minute (wpm), while the character spacing results in a much lower overall speed for the words.

By sending the characters at the higher speed, you keep from falling into the look-it-up-in-the-table trap. Thinking of the code as a table of letters in one column and a dot-dash pattern in another is a natural tendency. When you hear a sequence of code elements (the dots and dashes), such as short-long-long, you then look it up in your head to find the character "W." This method works, but only up to speeds of a few words per minute and it's a very hard habit to break.

The Farnsworth method gets you thinking "didahdah" instead of "dot dash dash." The former is closer to the actual sound of the character. You then memorize the character as the entire pattern of sounds and not as disconnected elements. This method keeps you from having to unlearn the table method and makes progressing to higher speeds easy. Make sure whatever study aid you choose uses the Farnsworth Method or something similar, such as ARRL and Gordon West Morse code study tapes and books or the *Ham University* and *Morse Academy* software. (Go to www.ac6v.com/morseprograms.htm for an encyclopedic listing of Morse code training aids.)

The graph in Figure 5-2 shows the normal progression in code speed. Between each step, or plateau, you achieve a new skill. While on the plateau, you refine or solidify the skill. Over time, you can progress from copying letter-by-letter to hearing whole groups of characters and then words.

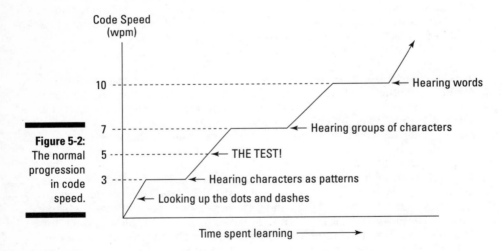

Figure 5-2:
The normal progression in code speed.

A great way to gauge your Morse code proficiency is with live code practice, which helps freshen up your taped or computer-generated material. The most widely received code practice sessions are transmitted by the ARRL's station W1AW from Newington, Connecticut, on several frequencies at different times throughout the day. W1AW transmits bulletins and code practice daily and through the weekends. You can find a complete W1AW operating schedule at `www.arrl.org/w1aw.html#w1awsked`. Code practice may be available on a VHF or UHF repeater in your area, too. Check with your local radio clubs to find out.

The most important part of mastering Morse code is to just keep at it. You'll have days where conquering new letters and higher speeds just seem to come effortlessly. Then days come when progress just seems elusive. Those plateaus are the most important time to keep going because that's when your brain is completing its new wiring!

As you discover more code, work it into everyday life. For example, while driving to work, whistle or hum the code for license plates, billboards, and signs. Just as you substitute foreign language equivalents for items you use or see as practice, use the code over and over again to make it familiar.

Soon, you'll notice yourself effortlessly copying bits and pieces that seemed impossible only days before. Characters that seemed hopelessly opaque become as natural as speech. Trust me — the 5 wpm you need to pass the test is within your grasp if you're willing to give it a try.

Chapter 6

Taking the Test

. . . .

In This Chapter
▶ Searching for a test session
▶ Registering for a test
▶ The Big Day

. . . .

*A*fter your diligent studies, you find yourself easily passing the online tests by a comfortable margin. Maybe you've been copying Morse code for a while and those 5-words-per-minute sessions are starting to seem a little slow. Now you're ready to (drumroll, please) . . . take the test!

If you are part of a class or study group, then the test may be part of the planned program. In this case, you're all set — just show up on time. Skip to the section, "The Big Day." If you are studying on your own, however, this chapter tells you where and when you can take the test.

Finding a Test Session

Lucky for you, finding a schedule of test sessions in your area is pretty easy. You can find the FCC Web site's list of organizations that serve as Volunteer Examiner Coordinators (VECs) in the different regions of the United States at wireless.fcc.gov/services/amateur/licensing/vecs.html#vecs.

If a VEC in the list is close to you, start by contacting the organization. Many of the VECs have a Web site and every one has an e-mail contact. Visit the Web site or send an e-mail that says, "Hello, my name is . . . and I want to take the Technician (or General or Amateur Extra) class license exam. Please send me a list of examination sites and dates. I live in. . . . "

If you don't see a nearby VEC or if no exams are scheduled at times or places suitable to you, you can find an exam with one of these VEC national organizations:

- ✔ **ARRL VEC exams:** The ARRL VEC operates nationwide and you can search for exams based on your zip code at `www.arrl.org/arrlvec/examsearch.phtml`, shown in Figure 6-1.

- ✔ **W5YI VEC exams:** Like the ARRL, the W5YI VEC (founded by Fred Maia W5YI) operates nationwide. You can find a list of certified examiners to contact at `www.w5yi.org/vol-exam.htm`.

- ✔ **W4VEC VEC exams:** The W4VEC (the call sign of the Volunteer Examiners Club of America) covers the Midwest and southern states and provides a list of dates and locations at `www.w4vec.com/ar.html`.

If you still can't find a convenient exam for you, your final option is to write or e-mail the VEC organizations at the addresses on the FCC Web site (`wireless.fcc.gov/services/amateur/licensing/vecs.html#vecs`) and ask for help. The mission of these organizations is to help prospective amateurs get a license. No matter where you live, they can put you in touch with examiners so that you can take your test.

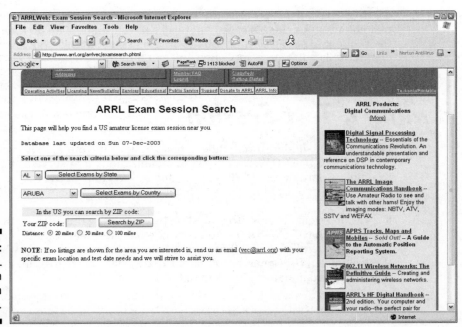

Figure 6-1:
The ARRL exam search page.

Signing Up for a Test

After you find a test session, you need to contact the test session hosts or sponsors to let them know that you are attending the session and what test elements you want to take. Checking in ahead of time is not only good manners, but can alert you to time or location changes. Some sessions do not allow unannounced attendees, or *walk-ins,* so be sure to contact the test sponsor first.

Public exams

Most exam sessions are open to the public and are held at schools, churches, or other public meeting places. Nearly all test sessions are open to walk-ins — that is, you can just show up unannounced, pay your test fee, and take the test — but some require an appointment or reservation. Checking first, before you show up, is always a good idea.

Call or e-mail the session's contact person to confirm the date, get directions, and let him or her know what tests you need to take.

Exams at events

Test sessions held at public events are very popular ways to take a test. Finding exam sessions being held at hamfests and conventions is very common. (See Chapter 3 for information on these events.) These sessions can attract dozens of examinees and often fill up quickly. Some exams are often given more than once throughout the day, so you can take more than one test or can spend time enjoying the event. As a test-taker, the FCC says you should not be required to pay for attendance at the event just to take license exams, but you may encounter a special entry fee. Don't be afraid to call ahead and ask!

If you attend an event-sponsored exam, getting to the site early and registering is a good idea. There may likely be multiple sessions and the tests for different elements may only be given at specific times.

Private exams

While the lion's share of tests are given at public exam sessions, smaller sessions may be held in a private residence, especially for a rural and small-town testing facility. For example, since 1990, most of the new hams in my community have passed their exams while sitting at my kitchen table!

When taking a test at a session in a private residence, please call ahead to ensure room for you at the session and the examiners can prepare to administer the exam you want to take. You are a guest in someone's home, so act accordingly.

The Big Day

In the so-called Good Old Days, the higher-class license exam sessions were conducted in federal office buildings by FCC employees. I vividly recall standing in line with dozens of other hams waiting for my shot at a new license. Some of us drove for hours to reach the FCC office or test location, nervously reviewing the material or listening to code tapes between swallows of coffee. Inside, a steely-eyed examiner watched over us as our pencils scratched out the answers.

These days, the exams are certainly more conveniently offered and the examiners friendlier, but that doesn't lessen your nervous anticipation as the day arrives. The best way to do well, of course, is to be prepared. That means in all aspects of the exams, not just the questions. The more you know, the less you have to worry about.

What to bring with you

In each test session, the three basic steps are

1. Register for your exam.
2. Take the test.
3. Complete your paperwork (which I talk about in Chapter 7).

The examiners guide you through each step.

When coming to a test session, be sure to bring the following, whether you're licensed or not:

- Two forms of identification, including at least one photo ID, such as a drivers license or employer's identity card
- Your Social Security number (SSN)

If you have a license and are taking an exam to upgrade to a higher class, you also need to bring

- ✔ Your current original license and a photocopy
- ✔ Any original CSCE you have and a photocopy

 CSCE is an acronym for Certificate of Successful Completion of Examination. This certificate is your record of having passed one or more examinations. If you have just passed the Technician exam (Element 2), to get on the air you have to wait for the FCC to grant you a call sign. For any other license changes, the CSCE allows you to operate immediately with your new privileges.

- ✔ If you have one, you can substitute your FCC Licensee Number (CORES or FRN) for your Social Security number. (See Chapter 7.)

Along with your identification and any documents, also bring a couple of pencils and a calculator. You aren't permitted to use any kind of online device or a computer during the written exam unless you have a disability. (And you need to first coordinate the use of supporting devices with the test administrators.) If you bring scratch paper, it needs to be completely blank. For copying code, if you plan to use a typewriter, keyboard, or computer, let the test session administrators know in advance.

When you arrive at the test session, you sign in with your name, address, and call sign, if you have one. The test administrators review any identification and documents you have. You pay the test fee (currently $12 at exams through the larger VECs) and then you are ready to take the exam!

Taking the written exam

The time for the test is finally here. You've shown up and you've signed in. Now what? When do you start? Well, that depends on how many people are signed up ahead of you and how many different types of tests are given. In a small session, you may start the test immediately. In a larger session, you may have to wait a while until your turn comes. Each written test takes from 15 to 45 minutes; code tests are shorter. The session may be organized so that everyone starts and stops together or the testing may be continuous. The examiner explains the process for your session.

As I discuss in Chapter 5, the written tests are multiple-choice tests. You receive a pamphlet containing the test questions and an answer sheet for recording your choices.

Follow these sure-fire tips to help you turn that tiger of a test into a pussycat by keeping your thinker in top shape:

- ✔ **Don't:** Take the test when you're hungry, sleepy, or thirsty.

- ✔ **Do:** Wear a couple layers of clothing to make yourself comfortable whatever the room temperature, and visit the restroom before the session starts.

- ✔ **Don't:** Drink extra coffee or tea, but **do** take a vitamin or drink fruit juice.

- ✔ **Do:** Follow the directions for completing the identification part of your answer sheet, even though you may want to start the test right away.

- ✔ **Do:** Study a question that seems really difficult, and then move on to the next one. When you come back to the question later, it may seem crystal clear.

- ✔ **Don't:** Guess, unless it's your only option. Generally, your first choice is always your best choice. Unless you are quite sure, don't change your answers.

- ✔ **Do:** Completely erase the wrong answer or indicate clearly that you made a change if you change any answers.

- ✔ **Do:** Remember to breathe and take a minute to stretch, roll your head, or flex your arms and legs.

- ✔ **Do:** Double-check your answers before handing in your test to make sure you marked the answer you wanted.

When you're done with your exam, follow the administrator's instructions for turning in your paper, sit back, and try to exhale! Depending on the size of the session, you may have to wait several minutes for the administrator to grade your paper. At lease three Volunteer Examiners (VEs) verify the grades on all exams. Written exams require a score of 75 percent or better to pass. You pass code exams by either copying 25 characters in a row (numbers and punctuation count as two characters) or by correctly answering eight of ten questions about the text that is sent.

In all probability, because you studied hard and seriously, you get a big smile and a thumbs-up from the test graders! Way to go! You can finally, truly relax and move on to the next stage.

If you didn't make it, don't be disheartened. Many sessions allow you to take a different version of the test as a second chance, if you want, both for code and for written exams. Even if you don't take the exam again right away, you now know the ropes of a test session and you'll be more relaxed next time. Don't let a failure stop you! Many hams make more than one attempt to pass a test and they are on the air today.

Passing the Morse code test

The Morse code test is very basic: You listen to five minutes of code from a tape or computer and write down (copy) what characters you hear. The test goes something like this:

1. **The VE starts the tape or computer and you get one minute of practice code before the real test begins.**

   Be sure that you can hear the code generator clearly and let the test administrator know if you can't. Sometimes strong echoes interfere with the code, so don't be afraid to have the equipment adjusted or move around until you can hear clearly. Take advantage of this time to relax and get comfortable. Don't worry if you don't copy the practice minute well, use the practice to get rid of your jitters.

2. **Verbal instructions are given for the actual test code by the VE or from the test tape.**

   Follow the instructions carefully despite the adrenaline coursing through your veins.

3. **Six *V* characters (meaning *Test*) are sent.**

4. **The actual test begins and continues for at least five minutes.**

   The code contains mostly text, including punctuation and numbers. The text used for the ARRL-VEC exam is a sample contact with call signs and typical information, such as signal report, location, operator's name, and type and power of equipment and antennas. In short, it's very representative of the type of code you hear on the air.

   For other code exams, the text may be anything from plain English to random groups of characters, so you may want to check with the examiners first to avoid surprises.

5. **The volunteer examiners (VEs) evaluate your test paper.**

   You can pass the test in one of two ways:

   > You capture one minute of *solid copy* — 25 characters in a row (numbers and punctuation count for two characters), excluding spaces. Most of the students that I examine pass the test this way.

   > Answer eight out of ten questions about the contact you overheard.

Like written exams, you can take some easy steps to make your Morse test easier.

✔ **Don't:** Practice at very high or very low speeds, stay just a little faster than the test speed.

✔ **Do:** Have at least one spare pen or pencil ready to pick up and use.

✔ **Don't:** Scribble out or erase any miscopied characters, just make a single line through them like ~~THIS~~.

✔ **Do:** Leave an underscore when you miss a character, like __ this, because you may recognize the word as you copy more letters. Making the underscore for each missed character keeps you in the rhythm until you get back on track.

✔ **Don't:** Panic when you miss a character. Take a deep breath and focus on an upcoming character.

Chapter 7

Obtaining Your License and Call Sign

· ·

In This Chapter

▶ Filling out your paperwork

▶ Watching for your call sign

▶ Choosing your own call sign

▶ Keeping your license valid

· ·

*A*fter you pass your exam, only a small matter of paperwork separates you from your new license. Your exam session volunteers help you complete everything correctly and even send your paperwork in to the FCC for you. You still need to understand what you're filling out, though; that's what I cover in this chapter. Fill your paperwork out correctly and you won't delay the process of getting your call sign.

Completing Your Licensing Paperwork

After you successfully complete the exam, you need to fill out two forms:

✔ The first form you fill out is the CSCE (Certificate of Successful Completion of Examination).

Figure 7-1 shows the ARRL VEC CSCE. The VEC and FCC use the CSCE as a check against the test session records. Your copy of the completed form is documentation of your test credit and is used to show credit for the exam you passed at any other test session before you receive your license or upgrade from the FCC. Keep your copy of the CSCE until the FCC sends you a new license or records the change in its database. You probably want to hang on to it as a record of your achievement.

VEC: American Radio Relay League/VEC
CERTIFICATE of SUCCESSFUL COMPLETION of EXAMINATION

Test Site
(city/state): _____

Test
Date: _____

CREDIT for ELEMENTS PASSED
You have passed the telegraphy and/or written element(s) indicated at right. You will be given credit for the appropriate examination element(s), for up to 365 days from the date shown at the top of this certificate, if you wish to upgrade your license class again while a newly-upgraded license application is pending with the FCC.

LICENSE UPGRADE NOTICE
If you also hold a valid FCC-issued Amateur radio license grant, this Certificate validates temporary operation with the operating privileges of your new operator class (see Sction 97.9[b] of the FCC's Rules) until you are granted the license for your new operator class, or for a period of 365 days from the test date stated above on this certificate, whichever comes first. **Note:** If you hold a current FCC-granted (codeless) Technician class operator license, and if this certificate indicates Element 1 credit, this certificate indefinitely permits you HF operating privileges as specified in Section 97.301(e) of the FCC rules. This document must be kept indefinitely with your Technician class operator license in order to use these privileges.

LICENSE STATUS INQUIRIES
You can find out if a new license or upgrade has been "granted" by the FCC. For on-line inquiries see the FCC Web at http://www.fcc.gov/wtb/uls ("License Search" tab), or see the ARRL Web at http://www.arrl.org/fcc/fcclook.php3; or by calling FCC toll free at 888-225-5322; or by calling the ARRL at 1-860-594-0300 during business hours. Allow 15 days from the test date before calling.

THIS CERTIFICATE IS NOT A LICENSE, PERMIT, OR ANY OTHER KIND OF OPERATING AUTHORITY IN AND OF ITSELF. THE ELEMENT CREDITS AND/OR OPERATING PRIVILEGES THAT MAY BE INDICATED IN THE LICENSE UPGRADE NOTICE ARE VALID FOR 365 DAYS FROM THE TEST DATE. THE HOLDER NAMED HEREON MUST ALSO HAVE BEEN GRANTED AN AMATEUR RADIO LICENSE ISSUED BY THE FCC TO OPERATE ON THE AIR.

The applicant named herein has presented the following valid exam element credit(s) in order to qualify for the license earned category indicated below:
Circle the **bold** text from one or more of these examples:
—for pre 3/21/87 Technicians circle **3/21/87 Tech-EL 1+3**;
—for pre 2/14/91 Technicans circle **2/14/91 Tech-El 1**;
—for lifetime Novice code credit circle **Novice-El 1**;
—for a valid or expired-less-than-5-years FCC Radiotele-graph license/permit circle **FCC Telegraph-EL 1**;

NOTE TO VE TEAM:
COMPLETELY CROSS OUT ALL BOXES BELOW THAT DO NOT APPLY TO THIS CANDIDATE.

EXAM ELEMENTS EARNED

passed 5 wpm code element 1

passed written element 2

passed written element 3

passed written element 4

NEW LICENSE CLASS EARNED

TECHNICIAN

TECHNICIAN w/HF

GENERAL

EXTRA

Candidate's signature _____

Candidate's name _____ Call sign _____ (if none, write none)

Address _____

City _____ State _____ ZIP _____

VE #1 _____
signature   call sign

VE #2 _____
signature   call sign

VE #3 _____
signature   call sign

Candidate's copy=white•ARRL/VEC's copy=pink•VE Team's copy=yellow

Figure 7-1:
The ARRL
VEC CSCE.

✔ The second form you fill out is the NCVEC Form 605.

NCVEC Form 605 allows the FCC to process your new license. Whenever you get a new license, upgrade to a higher class, renew your license, change your name or address, or pick a new call sign, you use this form. You can also submit name, address, or call sign changes directly to the FCC by mail or online.

The volunteer examiners running the session send in your CSCE and NCVEC Form 605 to the certifying VEC organization. Asking your examiners about the average wait before the FCC updates your information in its database is a good idea, but on average, you wait about seven to ten business days.

If you're upgrading an existing license, you can go home and use those new privileges right away. You have to add suffixes to your call sign to note that you qualified for your new privileges. When your new license comes in the mail or your new license class is recorded in the FCC database, you can drop the temporary suffix. These new suffixes are the following:

✔ A General license adds /AG to your call sign on Morse code, and slash AG or temporary AG on voice.

✔ An Amateur Extras license adds /AE to your call sign on Morse code, and slash AE or temporary AE on voice.

Don't forget that you're required to maintain your current mailing address on file with the FCC. If you move residences or if your address changes, keep the FCC database up to date. Mail sent to the address in the FCC database should get to you in ten days or less. Your license is good only for ten years, so you eventually have to fill out another Form 605 to renew it.

Finding Your New Call Sign

After you complete the test session and your paperwork is sent to the VEC, you can begin watching the FCC database for your new call sign to appear. After your call sign appears with your name beside it, you can get on the air even if you don't have the paper license.

The FCC has an online licensee information system called ULS (Universal Licensing System), shown in Figure 7-2. Each licensee has a *Federal Registration Number,* or FRN, that serves as identification within the FCC. I outline the process of registering for your own FRN in the section "Registering with the FCC Online."

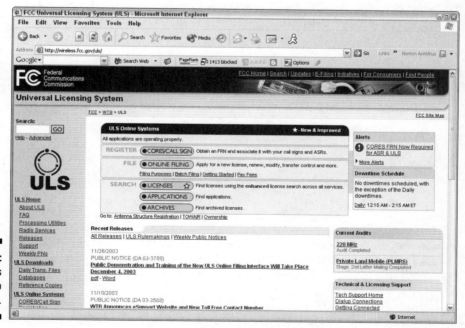

Figure 7-2:
The FCC's
ULS Web
page.

The database that really counts is the one maintained by the FCC. Follow these steps to find your call sign online:

1. **Log on to** `wireless2.fcc.gov/UlsApp/UlsSearch/`
   `searchLicense.jsp`.

   The FCC's Universal Licensing System (ULS) Web page loads (see Figure 7-3).

2. **Click the** <u>Amateur</u> **link in the Service-Specific column in the middle of the page.**

   The Search form appears.

3. **Enter your last name in the Name box and zip code in the Zip Code box in the Licensee section of the form, and then click the Submit button.**

   You may wait a few seconds for the information to come up while your request is processed. Figure 7-4 shows the results for **Silver** and **98070** — my whole family is licensed!

4. **Browse through the results.**

   If the results take up more than one page, click the <u>Query Download</u> link above the results to have the entire batch of results compiled into a single text file.

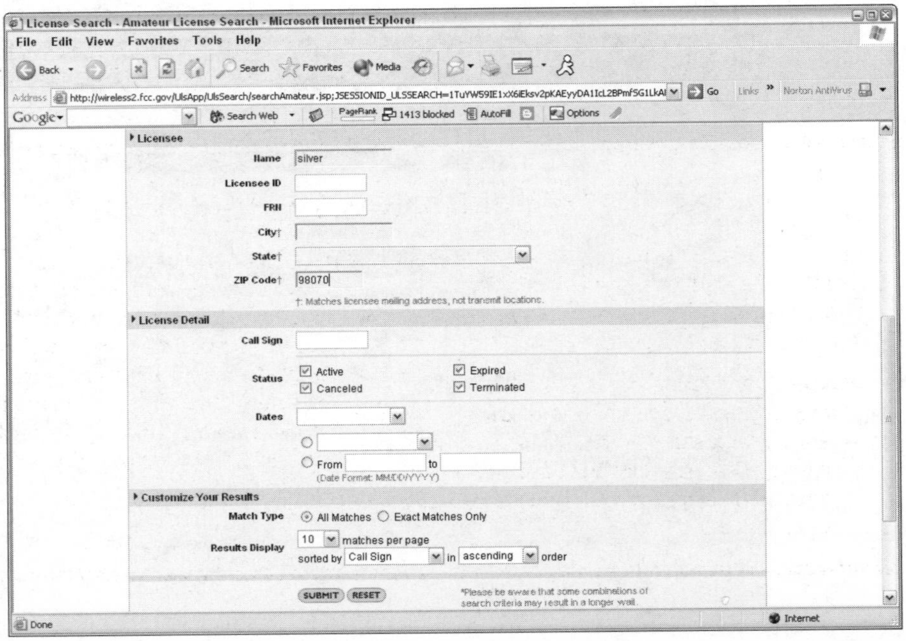

Figure 7-3:
The FCC's ULS Amateur License Search page.

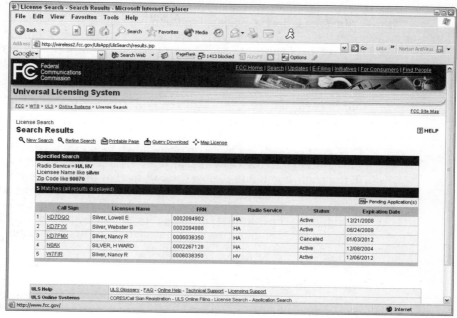

Figure 7-4:
The search results bring up everyone in my family.

Feel free to browse through the database. Seeing how many hams have the same last name as you, or how many are in your zip code is fun. By using a little creative investigating in the Amateur License Search page, you can discover some interesting things about the ham population in your area.

What if I don't see my call sign?

Patience is difficult while you're waiting, but be sure to wait for at least one full calendar week before getting worried. If two weeks pass, then you can take some action.

✔ Contact the leader of the exam session and ask if the VEC accepted your paperwork and sent it to the FCC. Due to some problem, a delay may have happened in getting the session results accepted. This is rare; don't worry.

✔ If the paperwork went through okay and it's been more than ten business days (or longer than the usual wait for the VEC that coordinated your session), ask the session leader to inquire about your paperwork. The VEC can trace all applications to the FCC. In more than ten years of being a session leader, I have never experienced any lost or delayed paperwork.

Most of the information you submit with your license application is a part of the FCC database and is available to the public along with your FRN, if you have one. When you submit your application, you agree to identify yourself as a licensed ham and state where your *fixed station* (a station that doesn't move) is located. The FCC does not make your Social Security number, phone number, or e-mail address available to the public.

In addition to the FCC Web site, you can also search other Web sites that search the FCC database:

- **QRZ.com** (www.qrz.com): The best-known ham radio Web site; just type the call sign into the Get Callsign search box at the top left-hand corner of the page. If you don't know the call sign, or prefer to search for a name, click the <u>Name Search</u> link in the Get Help area on the left. When the Name Search page loads, type your name, city, or zip code into the Search box. The information you get from this search comes directly from the FCC database. You can also search for non-U.S. hams if the data from their country is available online.

- **ARRL** (www.arrl.org/fcc/fcclook.php3): Type your last name and zip code and click the Submit Query button for results.

Registering with the FCC Online

The FCC has done a lot of work to make ordinary license transactions easier to accomplish by creating the online Universal Licensing System (ULS). The functions you can perform here include renewals, address changes, and other simple services. To use this system, though, you need to register in CORES, the Commission Registration System, whether or not the FCC has granted your license.

To register, you supply a TIN (Taxpayer Identification Number), which is your Social Security number, and you receive an FRN (Federal Registration Number) that is your personal identification with the FCC for any license you obtain. Each call sign is linked to a Licensee ID (the identification of the individual or organization that applied for the license). After you have your FRN, you can link your FRN with any call signs you hold.

Follow these steps to register with ULS/CORES as an individual Amateur licensee:

1. **Browse to** wireless.fcc.gov/uls/.

   The Universal License System Web page appears.

2. **Click the CORES/Call Sign button.**

3. **Select one of the following options and click the Continue button:**

   • **Register Now:** Select this option if this is your first time in the Universal Licensing System.

   • **Update Registration Information:** If you need to change any info from the last time you were in the system, select this option.

   • **Update Call Sign/ASR Information:** Choose this option if you need to add or remove any of your call signs from the list that you associated with your FRN. (Hams rarely use this option.)

4. **Select the An Individual option and the location of your contact address, and then click the Continue button.**

5. **Enter your name and address. Telephone and fax numbers and e-mail addresses are optional.**

   Any field marked with an asterisk is a required field.

   Everything you enter but your Social Security Number is available for public inspection. The FCC also keeps all telephone numbers and e-mail addresses private.

6. **If you are registering for the first time, enter your Social Security number.**

   You're required to provide this information (or give a reason why you can't). You must enter your SSN without any spaces, hyphens, or periods. Ignore any prompts or windows asking for an SGIN. (SGIN stands for a Sub-Group Identification Number and is used by managers of large communications services with many call signs. You don't need a SGIN.)

7. **At the bottom of the window, enter a password of 6 to 15 characters (or have the system pick one for you). Then re-enter it in the Re-enter Password box.**

   When selecting passwords, using your call sign in any way is not a good idea because an unauthorized person would try it as your password immediately. The same principle applies to your spouse's name, birth date, or any other personal information.

8. **Enter something that personally identifies you as being *you* in the Hint box.**

   If you ever forget your password and want the FCC to tell you what it is, this hint verifies you are who you are and not someone else. The personal identifier information could be your mother's maiden name or you can pick any word or words that fit in the window.

9. **Click the Submit button.**

10. **Correct any errors.**

    At this point, the ULS lists any errors you made, such as omitting a required item or adding the wrong type of information for a particular field. If any errors are listed, go back and correct them, and then click the Submit button again.

    The window now displays a form with your licensee information, as well as your Licensee ID Number, personal identifier, and password. Print this information and keep a copy in case you forget your password.

You are now a registered person with the FCC. Follow these steps to associate your call sign with your FRN:

1. **Click the FCC Universal Licensing System link at the bottom of the page that confirmed your information.**

2. **Click the Call Sign/ASR Registration link.**

    A window with your FRN already entered loads.

3. **Enter your password in the Password box and click the Continue button.**

4. **Click the Enter Call Signs link.**

    The Enter Call Signs window loads.

5. **Enter your call sign in the first space provided and click the Submit button.**

Congratulations! You and your call sign are now registered with the FCC. Numerous services are available to you for free!

Picking Your Own Call Sign

You can pick your very own call sign, within certain limits, of course. But if you're the sort of person who likes having a license plate that says "IMABOZO" or "UTURKEY," you'll enjoy picking a call sign from the available list.

Occasionally, you hear a call sign consisting of one letter, one numeral, and one number. These 1-by-1 call signs are granted on a temporary basis to U.S. hams for expeditions, conventions, public events, and all sorts of noteworthy activities. The special call sign program is administered by several of the VEC organizations for the FCC. More information is available at www.remote.arrl.org/arrlvec/1x1.html.

Ham radio license plates

You can also acquire a license plate with your call sign. The process is easy and many states even have a special type of vanity plate just for hams. Contact your local Department of Motor Vehicles and ask! For additional information see www.arrl.org/FandES/field/regulations/local/plates.html.

One wrinkle for hams with call signs that contain the Ø character: In most states, you have to request the slashed zero specially. Talk to the clerk that handles your form and show him or her your license. The slashed zero should be available for ham license plates.

Depending on your license class, you can select any available call sign in the groups listed in Table 7-1.

Table 7-1	Call Signs Available by License Class
License Class	*Types of Available Call Signs*
Technician and General	2x3; with a prefix of KA-KG, KI-KK, KM-KO, KR-KZ, and a suffix of any three letters
	1x3; with a prefix of K, N, or W, and a suffix of any three letters
Amateur Extra	2x3; with a prefix of KA-KG, KI-KK, KM-KO, or KR-KZ, and a suffix of any three letters
	2x1; with a prefix beginning with A, K, N, or W, and any letter in the suffix
	1x2; with a prefix of K, N, or W, and any two letters in the suffix
	1x3; with a prefix of K, N, or W, and a suffix of any three letters
Novice (no new licenses being issued)	2x3; with a prefix of KA-KG, KI-KK, KM-KO, or KR-KZ, and a suffix of any three letters
Advanced (no new licenses being issued)	2x2; with a prefix of K or W, and a suffix of any three letters

Needless to say, the shorter call signs and ones that seem to spell words are highly sought after. Many hams enjoy having a call made up of their initials. Whatever your preference, you'll likely find a vanity call sign that works for you.

You can find available call signs by using the FCC's ULS call sign search functions, but that can be quite cumbersome because it's designed to return information on only one call sign at a time. The following Web sites offer better and more flexible call sign search capabilities:

- ✔ **N4MC's Vanity HQ** (www.vanityhq.com/): By using N4MC's Web site, you can quickly determine whether call signs are available for assignment.

- ✔ **WM7D Callsign.net Callsign Database** (www.wm7d.net/fcc/callsign.html): This site includes a good search function that allows wildcard characters, which speeds up your search for that perfect call sign.

- ✔ **Vanity Call Sign Search** (www.amateur-radio.org/vanity.htm): This basic site has a good selection of search-and-sort features and links to other vanity call sign Web sites.

After you select a vanity call sign, you can then follow the vanity call application process described at www.arrl.org/arrlvec/vanity.html. You need to provide a list of one or more call signs that you like. All the call signs must be unassigned and available — that's what you use the vanity call Web sites for. Then you fill out the necessary application online and either pay the $16.30 fee via credit card or by check.

Maintaining Your License

The FCC's ULS has a specific section for the Amateur Radio Service that you can find at wireless.fcc.gov/services/amateur/licensing/filing.html. The ULS is the place to go if you want to do any of the following:

- ✔ Renew your license
- ✔ Change any of the address information associated with the license, such as name or address
- ✔ Replace your physical license
- ✔ Check on an application
- ✔ Apply for a vanity call sign

Checking out this page when you first earn your license is a good idea so that you are familiar with it when you need to take care of any licensing business.

Part III
Hamming It Up

The 5th Wave   By Rich Tennant

"Oh, him? He's some guy from Muncie, Indiana looking for
an Elmer to help log his first QSO."

In this part . . .

In this part, you find out about the real ham radio: making contacts. I start by showing you how (and where and when) to make contacts and what they usually consist of. I show you how to use those cool handheld radios that everybody seems to have as well as how to *pound brass* — ham speak for Morse code.

Ham radio also comes through when the chips are down in emergencies, natural disasters, and just providing public service. Definitely read these sections just to familiarize yourself with the emergency-related possibilities — even better, join in.

Many amazing activities take place on the ham bands every day, including round-the-world DX-ing, making contacts through amateur satellites, using your computer to send and receive data using a radio, taking place in on-the-air competitions, even having your own amateur TV station!

Chapter 8

Making Contact

*W*hen you have your ham radio *(rig)* set up and a license from the FCC *(ticket)* clearing you for takeoff, you're ready to make your first connection. If this thought makes your palms a little sweaty, don't worry; all hams start out feeling just that way and survive. You will, too.

In this chapter I show you how to make a contact and I cover the basics of on-the-air manners and the simple methods that make contacts flow smoothly. With a little preparation, you'll feel comfortable and confident, ready to get on the air and join the fun.

Listen, Listen, Listen!

The most important part of successfully putting a contact, also known as a *QSO,* in your logbook is listening. In fact, your ears are the most powerful part of your station! The ham bands are a 24-hour-a-day party with people coming and going all the time. And just like when you walk in to a big party, you need to size up the room by *tuning the band* (listening at different frequencies to assess activity) or *monitoring* (listening to an ongoing contact or conversation) for a while before jumping in. By listening, you discover who's out there and what they're doing, what the radio conditions are like, and the best way for you to make contact.

Listening on the different bands

Operating on the shortwave or HF bands has a different flavor from the VHF bands. On the HF bands, you can find stations on any frequency that offers them a clear spot for a contact. Up on the VHF bands, most contacts take place using repeaters on specific frequencies or channels spaced regularly by a few kHz. How are you supposed to figure out where the other hams hang out?

HF, or *High Frequency,* bands cover 3 to 30 MHz and are usually thought of as the shortwave bands. VHF, or *Very High Frequency,* bands cover from 30 to 300 MHz. UHF, or *Ultra High Frequency,* bands cover 300 MHz to 3 GHz.

Repeaters are radios that listen on one frequency and retransmit what they hear on another frequency. Repeaters are usually in high spots such as hilltops or on tall towers so they can hear weak signals well and they have powerful transmitters so that they can be heard for a long distance. Repeaters allow weak portable and mobile stations to communicate with each other over a wide area. Repeaters are most useful on VHF and higher frequency bands.

On both HF and VHF, hams engage in specific activities and tend to congregate on or near specific frequencies. For example, low-power (or *QRP*) aficionados are often on the 20-meter band near 14.060 MHz. No rule says they must operate on that frequency, but they gather there routinely anyway. Having that kind of consistency provides a convenient way to meet others with similar interests. To continue my party metaphor, it's the same as learning from another party-goer, "A group is talking about jazz at that table in the corner." Whenever groups tend to congregate at particular frequencies, those frequencies are known as *calling frequencies.*

Understanding sub-bands and band plans

In the United States, specific regulations exist about where each type of signal transmits in a given band. These signals are called *sub-bands.* Figure 8-1 shows the sub-bands for the 80-meter band. The American Radio Relay League (ARRL) offers a handy chart showing the sub-bands on its Web site at `www.arrl.org/ FandES/field/regulations/bands.html`.

Outside the United States, regulations are much less restrictive. For example, you'll probably hear Canadian and overseas hams having voice contacts in a part of the band where American hams don't have phone transmitting privileges. (Phone is an abbreviation for *radiotelephone,* which includes all voice modes of transmission.) How unfair! Because of the number of American hams, the FCC long ago decided that to maintain order, segregating the wide-bandwidth phone signals from narrow-bandwidth code and data signals is necessary. That's just the way it is. So close and yet so far!

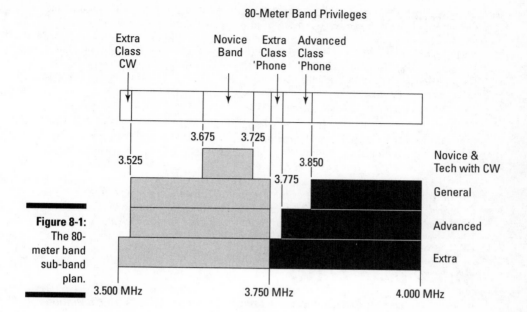

Figure 8-1:
The 80-
meter band
sub-band
plan.

Beyond this segregation of the amateur bands, amateurs have collectively organized themselves in order to organize the different operating styles on each
band. Not all amateur users can co-exist on the same frequency, so having
agreements about where the different types of operations occur is necessary.

Band plans are based on the FCC regulations, but go beyond them to recognize
popular calling frequencies and the segments where you can usually find certain operating styles or modes. A complete list of all band plans is beyond the
scope of this book, but a good source of up-to-date U.S. band plans is on the
ARRL Web site at www.arrl.org/FandES/field/regulations/bandplan.
html#17m. Be aware that different band plans are outside the United States.
Europe and Japan, for example, both have substantial differences on certain
bands.

Tuning In a Signal

Tuning in a signal consists of using the main tuning knob on the radio to change
the radio's operating frequency. The knob changes the frequency of a Variable
Frequency Oscillator, or VFO, that controls the radio's frequency. The radio's
operating frequency is displayed on a dial or display. A received signal's frequency is the radio's displayed frequency when the signal is tuned in properly.

The correct method of tuning in a signal depends on the type of signal it is. On signals that you receive or copy manually — such as Morse code, SSB, or FM — you use your ears. For signals sent using special equipment — such as RTTY, packet, or PSK31 — you have to use the displays from the equipment to get your receiver set just right. Whether you tune in a signal from above or below its frequency doesn't matter, although you may develop a preference for one or the other.

To start, set your radio to the type of signal you want to receive — CW, USB, LSB, or FM, for example. If your radio has a squelch control that mutes the audio unless a signal is present, you should set it to its off position to let all signals pass. That allows you to hear even a very weak signal as you tune by.

For help with the technical aspects of your equipment, try the *Ham Radio For Dummies* Web site (as described in the Introduction to this book). The next few sections outline a quick guide to tuning in the most common types of signals.

Morse code (CW)

Morse code signals are often referred to as CW, which stands for *Continuous Wave*. Early radio signals would die out quickly because they were generated by sparks. Soon hams discovered how to make steady signals, or continuous waves, that they can turn on and off with a telegraph key. Morse code and Continuous Wave (or CW) thus became synonymous.

To tune in to a Morse code signal, follow these steps:

1. **Set the rig to receive Morse code by selecting the CW mode.**

2. **Set the rig to use a wide filter if your rig has more than one filter available.**

   A wide filter allows you to find and tune in stations, while the narrower ones block out nearby signals. Selecting filters is done with a Wide/ Narrow control or with switches labeled with the filter's width. (Check the operating manual for precise instructions.)

3. **Adjust the main tuning dial until you hear a Morse code signal.**

   The pitch changes as you change the receiver's frequency. Tune until the pitch is comfortable to your ear. A low tone (300–600 Hz) is most restful to the ear, but a higher tone (500–1200 Hz) often sounds crisper. Most radios are designed so that when you tune in a signal with a tone or pitch around 500 Hz, the transmitted signal is heard by the other station at a similar pitch. If you prefer to listen to a note more than 100 Hz higher or lower, check your rig's operating manual to find out how you can adjust the radio to accommodate your preferred pitch.

4. **When you tune in the signal at your preferred pitch, select the narrower filter (if available) to reduce noise and interference.**

   If the frequency is not crowded or noisy, you can stay with the wider filter.

Single-sideband (SSB)

Single-sideband (SSB) is the most popular mode of voice transmission on the HF bands. (FM is mainly used above 50 MHz.) The mode got its name because of a key difference from the older mode, AM, which is used by AM broadcast stations and was the original voice mode hams used. Whereas an AM transmitter outputs two identical copies of the voice information, called sidebands, a SSB signal only outputs one. This signal is much more efficient and saves precious radio spectrum space.

Most voice signals on HF are SSB, so you have to choose between USB *(Upper Sideband)* and LSB *(Lower Sideband)*. The actual SSB signals extend in a narrow band above (USB) or below (LSB) the *carrier frequency* displayed on the radio (see Figure 8-2). How do you choose? By long tradition stemming from the design of the early sideband rigs, on the HF bands above 9 MHz, voice operation is always on USB. Below 9 MHz, you find everyone on LSB.

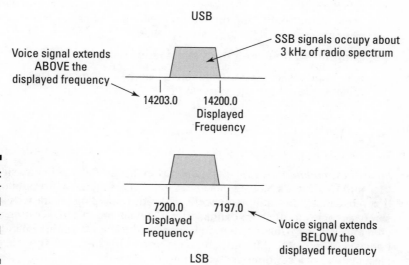

Figure 8-2: The Upper Sideband and Lower Sideband frequencies.

Because hams must keep all signals within the allocated bands, you need to remember where your signal is actually transmitted. Most voice signals occupy about 3 kHz of bandwidth. If the radio is set to USB, that means your signal appears on the air from the displayed frequency up to 3 kHz higher. Similarly on LSB, the signal appears up to 3 kHz below the displayed frequency. When operating close to the band edges, make sure your signal stays in the allocated band. For example, on 20-meters, the highest frequency allowed for hams is 14.350 MHz. You can tune a radio operating on USB no higher than 14.350 MHz -3 kHz = 14.347 MHz to stay legal.

If you tune across an old-style AM signal that has both upper and lower sidebands, you can tune in the signal using either USB or LSB. The whistling noise that gets lower and lower in frequency as you tune in the voice is the AM *carrier* frequency, centered precisely between the two voice sidebands. When the carrier tone goes so low that you can't hear it any more, you can listen to either sideband.

To tune into an SSB signal, follow these steps:

1. **Set your rig to receive SSB signals.**

   You may also have to choose Lower Sideband (LSB) or Upper Sideband (USB).

2. **Select the widest SSB filter.**

3. **Adjust the tuning dial until you hear the SSB frequency.**

   As you approach an SSB signal's frequency, you hear either high-pitched crackling (like duck quacking) or low-pitched rumbling. You can tell it's a voice from the rhythm, but it's unintelligible. These are the high or low frequency parts of the operator's voice.

4. **Continue to tune until the voice sounds natural.**

   If it sounds bass-y, your transmitted signal sounds too treble-y to the receiving operator and vice versa.

FM

Frequency modulation (FM) is the most popular mode of transmission on the VHF and UHF bands. FM signals encode the voice signal as frequency variations. Because hams adapted surplus FM radios used by businesses and public safety agencies to the ham bands, operation was organized as *channels* on specific frequencies. This convention is also used today, so tuning on most FM rigs consists of selecting different channels or moving between specific frequencies, instead of a continuous frequency adjustment.

To tune into an FM signal, follow these steps:

1. **Set your rig to operate on FM.**

   Most VHF/UHF radios only use FM, so your radio may have a control for selecting the mode.

2. **Set the squelch control so that noise is heard.**

   This process is called *opening* the squelch.

3. **Reset the squelch to stop the noise.**

   This process is to allow you to hear weak signals without having to listen to continuous noise. For very weak signals, you may have to re-open the squelch to receive it.

4. **Using a band plan or repeater directory as a guide, select a frequency.**

   If you are using an FM-only rig such as a hand-held or mobile unit, you can enter the frequency via a keypad, rotate a tuning dial that changes frequency as on HF, or select between different memory channels. If hams are active on that channel, you hear the operator's voice. Depending on the change in frequency with each step, you may have to tune back and forth to find the frequency where the voices sound best. If you are mis-tuned or *off frequency,* the voices are muffled or distorted.

   If you are using a multi-band radio that uses a main tuning dial, you can tune in the signal either by using your ear (tuning for the most natural-sounding voice with the least distortion) or, if available, by watching a tuning indicator called a *discriminator,* sometimes labeled DISC. The discriminator shows whether you are above or below the FM signal's center frequency. When the signal is centered, you're tuned just right.

Radioteletype (RTTY) and data signals

RTTY signals (sometimes pronounced *ritty*) consist of characters transmitted as a sequence of two different tones, called *mark* and *space*. Each character consists of a specific pattern of tones, alternating several times per second. RTTY signals sound like a warbling bird that only knows two notes. You cannot decode or generate RTTY signals manually, but you can with special equipment external to the radio that has a serial data connection to a computer, or by a computer sound card with special software. The radio treats the tones of a RTTY signal as a voice signal, either receiving or transmitting.

You can tune in data signals, such as PSK or MFSK, using very similar techniques and equipment. They are also constructed from audio tones and may sound like warbling or bursts of tones. In the following tuning steps, treat DATA as the equivalent to RTTY.

To tune into a RTTY signal, follow these steps:

1. **Set your radio to receive data signals, usually by a RTTY or DATA mode control.**

   If your radio doesn't have a DATA mode, use USB or LSB, following the USB/LSB convention for the band you're using. (The tones used for RTTY are near 2 kHz; selecting RTTY sometimes makes the audio noise from the receiver sound tinny as the receiver adjusts itself for higher-pitched tones.)

2. **Configure the external data encoder/decoder to provide a tuning indicator; if you have a sound card, open a tuning window.**

   You need to consult the operating manuals to determine how to do that. Tuning indicator styles vary, but most display a representation of the received tones with overlaid guide lines or figures that allow you to align the signal with the guide lines by tuning the receiver and adjusting the volume or AF Gain control.

3. **Tune in the data signal as you tune an SSB signal, but instead of listening for natural speech, watch the tuning indicator or window until the signal's tones properly align.**

   The received data window of the terminal or sound card software begins to show characters coming in. If the characters are garbled, be sure you are on the correct sideband (USB or LSB) or adjust the tuning slightly. Your software or decoder probably has a troubleshooting section to help you tune in the signal.

Listening on HF

Most of the traditional shortwave bands between 1.8 and 30 MHz are broadly organized into two segments. Digital transmissions such as CW (*Continuous Wave,* meaning, Morse code), radioteletype (or *RTTY*), and data occupy the lower segment. Voice signals occupy the higher segment. Within each of these segments, the lower frequencies are where you tend to find the long-distance (or *DX*) contacts, special-event stations, and contest operating. Casual conversations, known as *ragchews,* and scheduled on-the-air meetings *(nets)* generally take place on the higher frequencies within each band.

Table 8-1 provides some general guidelines where you can find the different types of activity. Depending on which activity holds your interest, start at one edge of the listed frequency ranges and start tuning. While tuning, use the widest filters your radio has for the mode (CW, SSB, or FM) that you select. You won't miss a station if you tune quickly, and finding the right frequency when you discover a contact is easier. After you tune in a contact, then you can tighten up your filters to narrower bandwidths, limiting what you hear to just the one contact.

Table 8-1	Activity Map for the HF Bands	
Band	*CW, RTTY, and Data Modes*	*Voice Modes*
160 Meters (1.8–2.0 MHz)	1.800–1.860 MHz (no fixed upper limit)	1.843–2.000 MHz
80 Meters (3.5–4.0 MHz)	3.500–3.750 MHz	3.750–4.000 MHz
60 Meters (5.5 MHz)	Not permitted	5.3305, 5.3465, 5.3665, 5.3715, and 5.4035 MHz
40 Meters (7.0–7.3 MHz)	7.000–7.150 MHz	7.150–7.300 MHz
30 Meters (10.1–10.15 MHz)	10.100–10.125 MHz CW 10.125–10.150 MHz RTTY and data	Not permitted
20 Meters (14.0–14.35 MHz)	14.000–14.150 MHz	14.150–14.350 MHz
17 Meters (18.068–18.168 MHz)	18.068–18.100 MHz (no fixed upper limit)	18.110–18.168 MHz
15 Meters (21.0–21.45 MHz)	21.000–21.200 MHz	21.200–21.450 MHz
12 Meters (24.89–24.99 MHz)	24.890–24.930 MHz (no fixed upper limit)	24.930–24.990 MHz
10 Meters (28.0–29.7 MHz)	28.000–28.300 MHz	28.300–29.7 MHz (most activity below 28.600 MHz)

If every voice you hear seems scrambled, your rig is probably set to receive the wrong sideband. Change sidebands and try tuning again.

Because the ionosphere strongly affects the signals on the HF bands as they go from point A to point B, the time of day makes a big difference. On the lower bands, the lower layers of the ionosphere absorb signals through the day, but disappear at night, allowing signals to reflect off the higher layers and reflect over long distances. Conversely, the higher bands require the sun's illumination for the layers to reflect HF signals back to Earth, supporting long-distance *hops* or *skips*. Table 8-2 shows the general guidelines for what you hear on the different HF bands at different times of day. With the exception of sporadic effects, the ionosphere is much less a factor on the VHF and UHF bands at 50 MHz and above.

Table 8-2	Day/Night HF Band Usage	
HF Band	*Day*	*Night*
160-, 80-, and 60-meters (1.8, 3.5, and 5 MHz)	Local and regional out to 100–200 miles.	Local to long distance with DX best near sunset or sunrise at one or both ends of the contact.
40- and 30-meters (7 and 10 MHz)	Local and regional out to 300–400 miles.	Short-range (20 or 30 miles) and medium distances (150 miles) to worldwide.
20- and 17-meters (14 and 18 MHz)	Regional to long distance. Bands open at or near sunrise and close at night.	20-meters is often open to the west at night and may be open 24 hours a day.
15-, 12-, and 10-meters (21, 24, and 28 MHz)	Primarily long distance (1,000 miles and more). Bands open to the east after sunrise and to the west in the afternoon.	10-meters is often used for local communications 24 hours a day.

Listening on VHF and UHF

Most contacts on the VHF and UHF bands are made using repeaters. Repeaters are most useful for local and regional communication, allowing individual hams to use low-power hand-held or mobile radios to make contacts over that same wide area. For this scheme to work, the repeater *input* and *output* frequencies are fixed and well-known, so the bands are organized into sets of channels. (Repeaters are not used on the HF bands because of the need to both receive and transmit simultaneously — difficult within single HF bands.)

Most VHF and UHF voice contacts use the FM or *frequency modulation* mode of voice transmission because of its excellent noise suppression, making for easy listening. The drawback is that FM doesn't have the range of CW or SSB transmissions. Contacts made directly between hams using FM are referred to as *simplex,* and using a repeater, as *duplex.*

Repeater and simplex FM channels are generally separated by 20 kHz. You can view a complete band plan for the 2-meter and 70-cm bands at www.arrl.org/ FandES/field/regulations/bands.html. I cover repeater operation in more detail in Chapter 9.

Dealing with beacons

After you tune the bands, you may still not know for sure whether the band is *open* (meaning that signals can travel beyond line-of-sight), and in what direction. Propagation software is available to offer predictions to help make those decisions, but those predictions are, well, only predictions. To help determine whether a band is open, beacon transmitters are set up around the world. A *beacon* is a transmitter that sends a message continuously on a known frequency. Amateurs receiving that beacon's signal know that the band is open to the beacon's location.

The biggest network of beacons in the world is run by the Northern California DX Foundation (NCDXF). The network consists of 18 beacons around the world, as shown in the following figure. These beacons transmit on the 20- through 10-meter bands in a round-robin sequence. They also vary their transmitting power from 100 to 1 watt so that hams receiving the beacon signal can judge the quality of propagation. A complete description of this useful network is available at `www.ncdxf.org/beacons`.

Other amateurs have set up their own beacons, too. You can find listings of frequencies for these amateur beacons on various Web sites. A good reference for all beacons on HF and 6-meters is the excellent list at `www.keele.ac.uk/depts/por/28.htm`. Amateur VHF beacons are listed on several Web sites; just enter **amateur VHF beacon** into a Web search engine to locate several beacon listings.

Repeaters enable you to use low-power and mobile radios to communicate over a large distance. Many hams use repeaters as a kind of intercom to keep in touch with friends and family members as they go about their daily business. These contacts are generally much less formal than on HF and you likely hear contacts between the same groups of hams every day. Repeaters are where you find local hams and find out about local events.

To make contact via a repeater, you may have to enable *tone-access* on your radio. Tone access adds one of several standard low-frequency tones to your speech audio to let the repeater know that your signal is intended for it and is not interference. If you do not transmit the required tone, the repeater does not re-transmit your signal and you can not be heard. (The radio's operating manual can tell you how to select and activate the tones.)

Not all repeater channels have an active repeater. In order to find repeaters in your area or while traveling, repeater directories are available from several sources. Some of these directories are nationwide; local clubs or repeater organizations publish others and focus on a specific region (see Figure 8-3).

The directories and Web sites list frequencies and locations of repeaters so you can tell which may be available in your area. Repeater directories also list the required access tones and other operating information and features for individual repeaters.

To listen to repeater contacts, follow these steps:

1. **Use a repeater directory or Web search to find a repeater in your area.**

2. **Determine the repeater's input and output frequencies.**

3. **Set up your radio to listen on the repeater's output frequency.**

    You can also listen to stations transmitting to the repeater, called *listening on the input*.

4. **Tune your radio as you do for FM signals.**

    If you need help tuning your radio, turn to the section "Tuning In a Signal," earlier in this chapter.

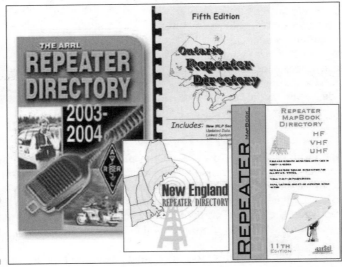

Figure 8-3:
Repeater
directories.

For direct ham-to-ham contacts on VHF and UHF over distances where FM results in noisy, unpleasant contacts, use the more efficient CW and SSB modes, which are called *weak signal* communication on VHF and UHF because you can make contacts with much lower signal levels than by using FM. The very lowest segments of the VHF and UHF bands are set aside for weak-signal operation. Weak-signal operations are conducted in much the same way as SSB and CW on HF with contacts taking place on semi-random frequencies centered around calling frequencies. To tune in contacts on the weak-signal segments of VHF and UHF bands, see the section "Tuning In a Signal," earlier in this chapter.

Deciphering a QSO

As you tune across the bands, dozens of contacts may be going on. That sounds like a bewildering variety, but you'll find that most contacts (or QSOs) are one of three types:

- ✔ Casual conversation *(ragchews)*
- ✔ Nets
- ✔ Contesting *(DX-ing)*

Chewing the rag

Chewing the rag is probably the oldest type of activity in ham radio. I have no idea where the expression came from, except possibly from "chewing the fat," but if you like to chat, you're a *ragchewer* and are following in the footsteps of the mythical master ragchewer The Old Sock, himself! Ragchewing is an excellent way to build your operating skills, perhaps leading to an award such as the A-1 Operator Club Award shown in Figure 8-4.

Figure 8-4:
The ARRL offers several operating awards, including the A-1 Operator Club shown here and the Code Proficiency Certificate.

The American Radio Relay League, Inc.

A-1 OPERATOR CLUB

This certifies that

H. Ward Silver, NØAX

is a member of the ARRL A-1 Operator Club and is authorized to nominate other deserving qualified radio amateurs for membership.

Membership in the A-1 Operator Club represents adherence to the several principles of good operating: (1) Careful keying and good voice operating practice; (2) Correct procedure; (3) Copying ability; (4) Judgement and courtesy.

04/26/2000

W6ROD
President

Keep these things in mind while you're chewing the rag:

- ✔ **Start with your basic information: your call sign, signal report, operator name, and the location of your station.** Ragchews may be conducted between hams in the same town or across the world from each other.

- ✔ **After you exchange basic information, you may wander off in any direction.** Hams talk about family, other hobbies, work, propagation, technical topics, operating, you name it, just about anything is discussed.

  In general, hams avoid talking about politics or religious topics and don't use profanity. That leaves a lot to talk about, and hams seem to cover most of it.

- ✔ **Wrap up the contact when you run out of things to talk about, conditions change, or maintaining contact is difficult.** Exchange call signs once more, and tune away.

Round tables are where a group of hams on one frequency share a contact informally. Each ham talks in turn and all get a chance in sequence. A moderator may also regulate round tables.

Meeting on nets

Nets, an abbreviation of *networks,* meet at a regular time and on a consistent frequency. They have a theme — message handling (or *traffic*), emergency communication training, nets for mariners or mobile stations, or topics such as antique radios or technical Q&A.

Follow these tips when accessing a net:

- ✔ **Check in with the net control station, or NCS, and list your business.** The NCS orchestrates all exchanges of information and formally terminates the net when business is concluded.

  The net may meet until all business is taken care of or just for a fixed amount of time.

- ✔ **If you're a visitor, find out the specific time you can check in.** Nets often have a specific time when visitor stations can check in.

  You can find nets on a specific topic or frequency online. For example, the ARRL Net Directory is at `www.remote.arrl.org/FandES/field/nets`.

I discuss nets and net operation in more detail in Chapter 10.

Contesting and chasing DX

Radio contests are competitions in which you exchange call signs and a short message as quickly as possible. *Chasing DX,* or *DX-ing,* is the pursuit of contacts with distant (for which DX stands) stations.

While contesting or DX-ing, follow these rules:

✔ **Keep contest contacts short.** In a contest, the object is generally to make the largest number of contacts, so dilly-dallying around is not desirable.

✔ **Pass along just the minimum amount of information, called the** *exchange.* Then sign off in search of more contacts.

While DX-ing, keep contacts with rare stations short — just your call sign and a signal report — if many are calling. Keeping contacts short allows other hams to make a desirable contact, too. Ragchews with DX stations are encouraged, if conditions support good signals in both directions. Try to judge conditions and tailor your contact appropriately.

DX contacts are short because the distances are great and maintaining contact is difficult.

✔ **If you encounter stations making contest QSOs, listen until you figure out what information is being exchanged before participating.** By far, the most common information is a signal report and your location (often expressed as a numbered zone or section defined by the contest sponsor) or *serial number* (assign a serial number for each contact you make in the contest). If it's your fifth QSO in that contest, your serial number is 5, for example. The contesters are happy to explain what information they need. You can usually find the complete rules for contests in magazines or on the sponsor's Web site.

✔ **Remember that not everyone speaks English.** Most hams often know enough words in English to communicate a name, location, and signal report. Otherwise, an international set of Q-signals allows you to exchange a lot of useful information with someone with whom you don't share a common language.

See the next section, "Q-Signals," for more on Q-signals. DX-ing is a great way to exercise that rusty high-school Spanish or German, too!

I cover contests and DX-ing in more detail in Chapter 12.

Q-Signals

Q-signals began in the early days of radio as a set of standard abbreviations to save time and allow radio operators who did not share a common language to communicate effectively. Today, amateurs use Q-signals as shorthand to speed up communication. The definitions have drifted a little over the near-century of radio, but you find their use ubiquitous. (Table 8-3 lists many common Q-signals.)

During contacts, Q-signals often take the form of a question. For example, QTH? means "What is your location?" The reply, QTH New York, means "My location is New York."

Table 8-3	Common Q-Signals	
Q-Signal	**Meaning As a Query**	**Meaning As a Response**
QRG	What is my exact frequency?	Your exact frequency is _____ kHz.
QRL	Is the frequency busy?	The frequency is busy. Please do not interfere.
QRM	Are you being interfered with?	I am being interfered with. (As a noun: interference)
QRN	Are you receiving static?	I am receiving static. (As a noun: static)
QRO	Shall I increase power?	Increase power.
QRP	Shall I decrease power?	Decrease power.
QRQ	Shall I send faster?	Send faster (__WPM).
QRS	Shall I send more slowly?	Send more slowly (__WPM).
QRT	Shall I stop sending?	Stop sending.
QRU	Have you anything more for me?	I have nothing more for you.
QRV	Are you ready?	I am ready.
QRX	Do you want me to standby?	Standby.
QRZ	Who is calling me?	Not used as a response.
QSB	Is my signal fading?	Your signal is fading.

Q-Signal	Meaning As a Query	Meaning As a Response
QSK	Refers to "break in operation" where the sending station can receive between Morse code characters or individual dits and dahs.	
QSL	Did you receive and understand my transmission?	Your transmission was received and understood.
QSO	Abbreviation for a contact.	
QSP	Can you relay to___?	I can relay to____.
QST	General call preceding a message addressed to all amateurs.	
QSX	Can you receive on ___ kHz?	I am listening on ___ kHz.
QSY	Can you change to transmit on another frequency (or to ___ kHz).	I can change to transmit on another frequency (or to ___ kHz).
QTC	Do you have messages for me?	I have messages for you.
QTH	What is your location?	My location is ____.

Making a Call

Your big moment approaches! In this example, I use the dummy call sign of KD7FYX. Replace this call sign with your own. Follow these steps when tuning one of the HF bands:

1. **Find someone to talk to.**

   When you come across a fellow ham making a general "come in, anybody" call, you have found someone *calling CQ.* This situation is the easiest way for you to make a contact. You hear something like this: *"CQ CQ CQ this is November Zero Alpha X-ray standing by. . . ."*

   November, Alpha, and X-ray are *phonetics* that represent the letters of my call sign, NØAX. Phonetics are used because many letters sound the same (think B, E, T, P) and the words help get the exact call sign across. Table 8-4 lists the standard phonetics hams use. You may encounter alternatives, such as *Germany* instead of *Golf,* for instance. When in doubt, I respond or call with the phonetics used by the station I want to contact.

2. **Carefully note their call sign and respond.**

   Press the microphone button, and say: *"November Zero Alpha X-ray this is Kilo Delta Seven Foxtrot Yankee X-ray (repeat twice more), over."*

   Give the calling station's call sign once (you don't have to repeat this — they already know it!) and then give yours three times (a *1-by-3* call). If the calling station is very strong, you may just give your call twice. You don't need to repeat their call sign or shout, just speak in a normal, clear voice.

3. **Listen for the response if the station hears you.**

   You may hear something like: *"KD7FYX (possibly in phonetics) from NØAX, thanks for the call. Your signal report is. . . ."*

And you have a QSO in your logbook!

Making this kind of contact works a little differently on a repeater. Hams mostly use repeaters as a kind of regional intercom, so they are less likely to make a general call for a random QSO. For example, you never hear *CQ CQ CQ . . .* on a repeater. Hams turn the radio on in the car or in the shack to listen for friends or just monitor the frequencies. Your cue that someone is available for a contact is that they announce their presence by saying, *"This is NØAX, monitoring,"* or some other kind of general "I'm here" announcement. Just give them a 1-by-1 call, *"NØAX this is KD7FYX,"* and see if they come back to you. You also get good results by calling a station immediately after they complete a contact. Repeater usage tends to be more utility-oriented than on HF and you may find the contacts a little briefer. To respond on a repeater where signals are probably quite clear, just give a 1-by-1 call: *"NØAX this is KD7FYX."*

On CW or the digital modes, the process is much the same and looks like this:

1. **Copy the calling station's call sign.**

   You hear something like: *"CQ CQ CQ DE NØAX NØAX NØAX K."*

   DE is the telegrapher's shorthand for "From." K means "End of Transmission, Go Ahead." (**Note:** No upper- or lowercase characters are in Morse code — "de" is equivalent to "DE.")

2. **Respond with a 1-by-2 or 1-by-3 call.**

   Say something like: *"NØAX DE KD7FYX KD7FYX KD7FYX K."*

3. **Listen for a response.**

   You probably hear something like: *"KD7FYX DE NØAX TKS FOR THE CALL—UR RST. . . ."*

   TKS is shorthand for "Thanks," and UR is shorthand for "Your."

Telegraphers and typists are a lazy lot and tend to use all sorts of abbreviations to shorten the text. A table of abbreviations is available at ac6v.com/morseaids.htm#CW.

Table 8-4	International Telecommunications Union Standard		
Letter	**Phonetic**	**Letter**	**Phonetic**
A	Alfa	N	November
B	Bravo	O	Oscar
C	Charlie	P	Papa
D	Delta	Q	Quebec
E	Echo	R	Romeo
F	Foxtrot	S	Sierra
G	Golf	T	Tango
H	Hotel	U	Uniform
I	India	V	Victor
J	Juliet	W	Whiskey
K	Kilo	X	X-ray
L	Lima	Y	Yankee
M	Mike	Z	Zulu

Failing to make contact

But what if you try to make a contact and your call doesn't get answered with a response? Your signal may be too weak to hear or the station may have strong noise or interference that you can't hear. In this case, you find another station to call. Assuming that your signal is strong enough for other stations to hear it however, several other things may have happened:

✔ **Other hams are calling at the same time.** You can either wait around until the station you intended to contact is free and try again, or you can tune around for another contact opportunity. The most important thing is not to get discouraged!

✔ **The calling station can hear you, but can't make out your call.** The ham may either ask you to call again or respond to you, but won't have your call sign correct. The station may say: *"station calling, please come again."* Or *"QRZed?"* or *"Who is the station calling?"*

QRZed? is the international Q-signal for "Who is calling me?" Hams often use the British pronunciation, Zed, for the letter Z. At this point, just repeat your call two or three times, using standard phonetics and say, *Over* when you finish.

The station gets your call wrong by a letter or two. First stand by a few seconds to be sure another station with that call sign isn't on the same frequency. For example, I'm often on the air at the same time as NØAXL and NØXA and we are always getting confused. If a couple seconds go by and you don't hear another station responding, then respond with: *"NØAX this is KD7FYX (repeat twice more), do you have my call correct? Over."*

If repeated attempts at making contacts aren't producing results, check out your equipment. The easiest way is to locate a licensed friend and have him or her make a contact with you. That way, you know your transmitter is (or isn't) working and your signal is understandable. If you don't have a licensed friend, run through the following checklist to be sure you're transmitting what, when, and where you think you should be:

✔ **Are you sure you're transmitting on the right frequency?** Press the microphone switch or press the Morse key, and watch the radio's display very carefully. The indicators for frequency and sideband should stay exactly the same. You may be transmitting on a different sideband or frequency than you think you are!

✔ **Are you sure that your transmitter is producing power?** Watch the rig's power output meter to be sure the output power varies along with your voice or keying.

✔ **Is the antenna connected properly?** You should be receiving signals that are moderate to strong, indicating 4 to 9 S-units on the radio's meter. If the signals are very weak, you may have an antenna or cable problem. This problem also shows up as an SWR reading of more than 5:1 on your rig's SWR meter.

After your call is received correctly by the other station, proceed with the rest of the contact. Failed contacts and errors are handled very similarly on CW and the digital modes. Don't be bashful about correcting your call sign. After all, it's your radio name.

Breaking in is not hard to do

Sometimes you can't wait for the end of a contact to call a station. Interrupting another contact is called *breaking in*, or *breaking*, because the proper procedure is to wait for a pause in the contact and quickly say, *"Break"* (or send BK with Morse code).

Why do you want to do this? Perhaps you have an emergency and need to make contact right away. More frequently, you tune into a contact and the participants are talking about a topic or person that you know about. If you wait for the contact to end, you may not be able to contribute or help.

To break in a contact, follow these steps:

1. **Listen for a good opportunity to make your presence known.**

   When the stations switch transmitting and receiving roles is usually a good time to break in. You hear something like: *"So Lowell, back to you. KD7DQO from NØAX."*

2. **Quickly make a short transmission.**

   Don't be shy and wait or the other station begins transmitting. Say: *"Break."*

3. **Wait to see if either station heard your transmission.**

   One station hears you and says: *"This is KD7DQO, who's the breaker?"*

   If no one hears your transmission, start over with Step 1.

4. **Respond as if you're answering a CQ.**

   Say: *"KD7DQO this is KD7FYX (repeated), over."* Depending on the circumstances, you can then tell them your name and location before proceeding to explain why you broke in. At that point, the stations probably engage you in further conversation and you're in a three-way QSO! Sometimes, they won't want to have a third party in the contact, in which case you just courteously sign off and go on to the next contact.

Having a QSO

Because you listen to contacts (QSOs) on the air, you understand the general flow of the contacts. What do hams talk about, anyway? Like most casual contacts with a person you don't know, warming up to a contact takes a little time.

During the initial phase of the contact, you exchange information about the quality of the signals, your name, and your location. This phase is a friendly way of judging whether conditions permit you to have an extended contact. Then follow with information about your station and probably the local weather conditions. This information gives the other station an idea of your capabilities and whether static or noise is likely to be a problem.

Here are the common items you exchange when making a contact:

- **Signal Report:** This report is an indication of your signal's strength and clarity at the receiving station.

  - **SSB:** A two-digit system communicates readability and strength, although sometimes you can just use a single Q-number as a quality indicator.

  - **CW and RTTY:** The same two-digit system is used for readability and strength, but a third digit is added to indicate purity of the transmitter's note — rarely anything but 9 nowadays because transmitting equipment is quite good. If a poor signal is encountered, don't hesitate to give an appropriate report.

  - **FM:** The signal report is the degree to which the noise is covered up or *quieting*.

  - **Digital modes:** Use the same method as CW or RTTY, or don't bother with a signal report at all.

  See Table 8-5 for how to report your signal quality.

- **QTH (Location):** On HF, where signals take place over long distances, you generally give your town and state or province. You can give an actual address if requested, but it is usually not needed (if you aren't comfortable doing so). On VHF/UHF, you report the actual physical location, particularly if you're using a mobile radio.

- **Rig:** You can just report the power output shown on your transmitter's power meter (25 watts) or give the full model number and let your contact assume the transmitter is running at full output power.

- **Antenna:** Typically, you just report the style and number of elements, such as a two-element quad or ⅝-wave whip. Sometimes you can report a specific model number.

- **Weather:** Remember that stations outside the United States report the temperature in degrees Celsius! Standard weather abbreviations you can use on CW and digital modes include SNY, CLDY, OVRCST, RNY, and SNW. A Russian ham in Siberia once gave me a weather report of "VY SNW" (very snow).

Table 8-5	Reporting Signal Quality	
Mode	*System*	*Report Definitions*
SSB	RS: Readability & Strength	R is a value from 1 to 5; 5 means easy to understand, 3 is very difficult, 1 and 2 are rarely used.
		S is a value from 1 to 9. This number generally corresponds to the rig's signal strength meter reading on voice peaks.
	Q-(number): Indicates overall quality	Q1 to Q5; Q5 indicates excellent readability; reports below Q3 are rare.
CW	RST: Readability, Strength, and Tone	R is a value from 1 to 5, same as SSB.
		S is a value from 1 to 9, same as SSB.
		T is a value from 1 to 9; 9 being a pure tone and 1 being raspy noise. The letter C is sometimes added to indicate a chirpy signal.
FM	Level of Quieting (Signal report is for the station calling, not the repeater's output signal strength)	*Full Quieting* means almost all noise is suppressed. *Scratchy* means noise is present enough to partially disrupt understanding. *Flutter* means rapid variations in strength as a vehicle is moving. *Just Making It* means only strong enough to actuate the repeater, but not good enough for a contact.

After you go through the first stages of the QSO and the other station sounds like the ham wants to continue, you can try discussing some other personal information — your age, other hobbies, what you do for a living, your family members, and of course ham radio topics such as propagation conditions, any special interests, or particularly good contacts you made recently.

The FCC forbids obscene speech (which is pretty rare on the air). The three topics that seem to lead to elevated blood pressures are politics, religion, and sex — hardly surprising. So hams tend to find other things to talk about. Oh sure, you find some arguments on the air from time to time, just like in any group of people. Don't be drawn in to doing it yourself — no one benefits. Just spin the big knob and tune on by.

The trick is to keep your transmissions short enough so that the other station has a chance to respond or so that someone else can break in. That way, you both can have a QSO just as long and detailed as you want. If propagation is changing or the band is crowded or noisy, short transmissions allow you to ask for missed information. At the conclusion of the contact, you may also encourage the other station to call in again. Lifelong friendships are sometimes forged!

Calling CQ

After you make a few contacts, the lure of fame and fortune becomes too strong to resist — it's time to call CQ yourself! How do you go about it so that you sound right? If by now you have listened to dozens of CQs, which of them sounded the best to you? What makes them sound good?

A CQ consists of two basic parts repeated in a cycle. The first part is the CQ itself. For a general-purpose, "Hello, World!" just say *"CQ"*. If you're looking for a specific area or type of caller, you must add that information as in, *"CQ DX"* or *"CQ New England."*

The second part of the CQ is your call sign. You must speak or send clearly and correctly. Many stations mumble or rush through their call signs or send it differently each time, running the letters together. You probably tuned past CQs like that.

A few CQs followed by *"from,"* or *"DE,"* and a couple of call signs makes up the CQ cycle. If you say *"CQ"* three times followed by your call sign twice, that's a 3-by-2 call. If you repeat that pattern four times, it's a 3-by-2-by-4. At the end of the cycle, you add *"Standing by for a call,"* or *"K,"* to let everybody listening know that it's time to call.

An example of a 3-by-3-by-3 on CW is *"CQ CQ CQ DE NØAX NØAX NØAX CQ CQ CQ DE NØAX NØAX NØAX CQ CQ CQ DE NØAX NØAX NØAX K."*

Depending on conditions, repeat the cycle (CQ and your call sign) two to five times, keeping it consistent throughout. If the band is busy, keep it short. If you're calling on a quiet band or for a specific target area, four or five cycles may be required. When you're done, listen for at least a few seconds before starting a new cycle again to give anyone time to start transmitting.

Here's a short list of CQ Do's and Don'ts:

- **Do:** Keep your two-part cycle short to keep the caller's interest.
- **Do:** Use standard phonetics for your call sign on voice modes at least once per cycle.

- ✔ **Do:** Send CW at a speed you feel comfortable receiving.
- ✔ **Do:** Strive to sound friendly and enthusiastic.
- ✔ **Do:** Wait long enough between CQs for callers to answer.
- ✔ **Don't:** Mumble, rush, or slur your words.
- ✔ **Don't:** Send erratically or run letters together.
- ✔ **Don't:** Drag the CQ out. A 3-by-3-by-3 call is good for most conditions.
- ✔ **Don't:** Shout or turn up the microphone audio level too far. Clean audio sounds best.

Treat each CQ like a short advertisement for you and your station. It should make the listener think, "Yeah, I'd like to give this station a call!"

The long goodbye

If hams do one thing well, it's saying goodbye. Hams use abbreviations, friendly names and phrases, and colloquialisms that pad the contact before actually signing off. Hearing anyone say, *"Well, I don't have anything more to say, W1XYZ signing off,"* is rare.

Towards the end of the contact, let the other station know that you're out of gas. Some good endings are the following:

- ✔ **I AM** *QRU:* In Morse code speak, QRO means "out of things to talk about."
- ✔ **See you down the log:** Encourage another contact at a later time.
- ✔ **BCNU:** Morse code for "be seein' you."
- ✔ **CUL:** Morse code for "see you later."
- ✔ **88 or Old Man:** Throw love and kisses to any female operator, regardless of whether you know her well or not. Or tack on an affectionate "Old Man" (if appropriate).
- ✔ **73:** Don't forget your best regards! That's required!
- ✔ **Pulling the big switch or GOING QRT:** If you're leaving the airwaves, then be sure to say so after your call sign on the last transmission. On voice, say *"NØAX closing down"* or on a Morse code contact, *"NØAX SK CL."* Anyone listening knows you're vacating the frequency.

Sometimes, the signing off takes as long as the signing on.

Chapter 9

Casual Operating

· ·

In This Chapter

▶ Transmitting on repeaters

▶ Maintaining a casual contact

▶ Copying Morse code

▶ Picking up remote messages

· ·

*A*fter you tune around the amateur bands for a while, you'll agree that the lion's share of the ham's life is making relaxed, casual contacts. Some contacts are just random "Hello, anybody out there?" calls. You'll also hear contacts between hams that are obviously old friends or family members that meet on the air on a regular basis.

Lots of people also use ham radio to maintain contact, so to speak, while they're traveling. (I realize that this may be a shock, but in some places on Earth, a mobile phone's service indicator won't light up!) In remote places, ham radio is available to fill the breach. It's fun and people enjoy contacting travelers wherever they are.

I cover the technical aspects of station configuration and operation in Part IV. (For more technical stuff, see the *Ham Radio For Dummies* Web site.) In this chapter, you find out about the different ways to conduct these relaxing contacts. Like most things in life, with making casual contacts, "there's kindy a knack to it," as my dear Aunt Lexie used to say.

Before I start on this operating business, though, allow me to suggest that you get a copy of the FCC rules. You can download them from the FCC Web site or bookmark the URL (for reference), but the rules you find there, as written, are not altogether easy to figure out. For a clearer alternative, you can get *The FCC Rule Book* from the ARRL (www.arrl.org) for only a few dollars.

This rule book conveniently includes not only the rules themselves, but also a clear discussion on do's and don'ts, and information on technical standards and the FCC Universal Licensing System. New hams should really have a copy in their shack.

Operating on FM and Repeaters

Most new hams start out as Technician class licensees, which gives you access to the entire amateur VHF and UHF bands. By far the most common means of communicating on those bands is through the use of an FM repeater (see Chapter 8).

Finding a repeater

Figure 9-1 explains the general idea behind a repeater. A repeater retransmits (or *repeats*) FM signals that it hears on one frequency and simultaneously retransmits them on another frequency. The received signals are not stored and played back; they're re-broadcast on a different frequency at the same time they're received, which is called *duplex operation*.

Repeaters use FM instead of SSB because of the relative simplicity of the transmitters and receivers — an important consideration for equipment that is operating all the time and needs to be reliable. FM is also relatively immune from static if signal strength is good and so makes for a more pleasant contact. Except for a small segment of the 10-meter band, FM is rarely found on HF due to restrictions on signal bandwidth and FM's relatively poor quality on weak signals compared to SSB and CW. Above 30 MHz however, FM's qualities are ideal for local and regional coverage.

If the repeater is located on a high building, tower, or hill, its sensitive receiver picks up signals clearly from even tiny, handheld radios. It then uses a powerful transmitter to relay that input signal over a wide area. Stations can be separated by tens of miles, yet communicate with each other using just a watt or two of power by using a repeater.

Ham radio repeaters are constructed and maintained by either a radio club or private individual as a service to the ham community. If one exists in your area, consider joining or supporting a repeater user's group or club in order to help defray the cost of keeping the repeater on the air.

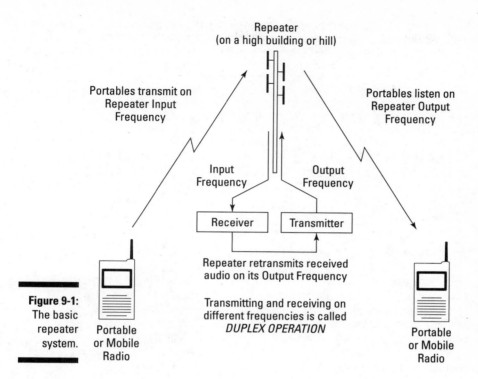

Repeater
(on a high building or hill)

Portables transmit on
Repeater Input
Frequency

Portables listen on
Repeater Output
Frequency

Input
Frequency

Output
Frequency

Receiver   Transmitter

Repeater retransmits received
audio on its Output Frequency

Transmitting and receiving on
different frequencies is called
DUPLEX OPERATION

Figure 9-1:
The basic
repeater
system.

Portable
or Mobile
Radio

Portable
or Mobile
Radio

To make repeater communications work, you have to know the frequency
on which a repeater listens and where it is transmitting. The listening frequency (the one that listens for your signal) is called the repeater's *input*
frequency and the frequency that you listen to is called the repeater's *output*
frequency. The difference between the two frequencies is called the repeater's
separation or *offset*. The combination of a repeater's input and output frequencies is called a *repeater pair*.

As Figure 9-2 shows, the repeater pairs are organized in groups with their
inputs in one part of the band and their outputs in another, all of them having
a common separation. Each pair leapfrogs its neighbor, with each input or
output *channel* separated by the same frequency, the *channel spacing*. The
input may be a lower frequency than the output or vice versa.

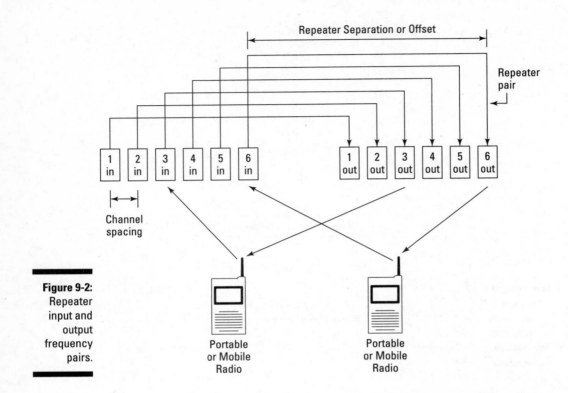

Figure 9-2:
Repeater
input and
output
frequency
pairs.

Figure 9-3 shows where the repeater segments of the five primary VHF/UHF bands are located. The 6-meter band has three groups of repeaters: 51.12 to 51.98 MHz, 52 to 53 MHz, and 53 to 54 MHz. The 2-meter band also has three groups: 144.6 to 145.5 MHz, 146 to 147 MHz, and 147 to 148 MHz. You can find a single group on the 222, 440, and 1296 MHz bands. Repeaters are allowed on the 902 MHz and 2304 MHz bands, but are not common. If you have a license with HF privileges, you may want to give the 10-meter FM repeaters a try with output frequencies between 29.610 and 29.700 MHz and an offset of –100 kHz.

Not all channels are occupied in every area and some variations are around the country as to channel spacing and, in rare cases, offset. In order to find out where the repeater inputs and outputs are for a specific area, you need a repeater directory (see Chapter 8 for more on repeater directories). The online directory at www.artscipub.com/repeaters/welcome.html is a good resource to locate nearby repeaters.

If you don't have a repeater directory and are just tuning across the band, try using the following table. Table 9-1 lists the most common output frequencies and repeater offsets to try. Tune to different output frequencies and listen for activity. ***Remember:*** You have to set your radio's offset appropriately for each band.

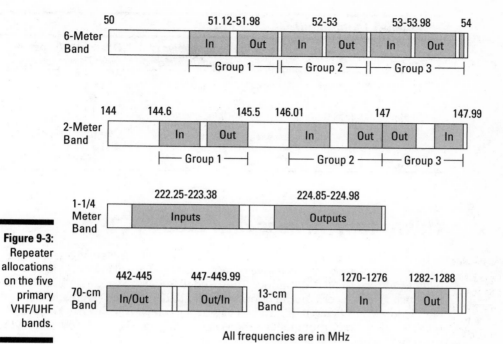

Figure 9-3: Repeater allocations on the five primary VHF/UHF bands.

All frequencies are in MHz

Table 9-1	Repeater Channel Spacings and Offsets	
Band	*Output Frequencies of Each Group (In MHz)*	*Offset from Output to Input Frequency*
6-meters	51.62–51.98	– 500 kHz
	52.5–52.98	
	53.5–53.98	
2-meters (a mix of 20 and 15 kHz channel spacing)	145.2–145.5	– 600 kHz
	146.61–147.00	– 600 kHz
	147.00–147.39	+ 600 kHz
220 MHz	223.85–224.98	– 1.6 MHz

(continued)

Table 9-1 *(continued)*

Band	Output Frequencies of Each Group (In MHz)	Offset from Output to Input Frequency
440 MHz (local options determine whether inputs are above or below outputs)	442–445 (California repeaters start at 440 MHz)	+ 5 MHz
	447–450	– 5 MHz
1296 MHz	1282–1288	– 12 MHz

Using tone access and DCS

In order to minimize interference from other repeaters and strong nearby signals, most repeaters now use *tone-access.* The tones are also known as *sub-audible, PL,* or *CTCSS* (Continuous Tone Coded Squelch System). You may have used tone-access on the popular Family Radio Service (FRS) and General Mobile Radio Service (GMRS) radios as *privacy codes.* Tone-access keeps a repeater or radio output quiet or *squelched* for all signals except ones carrying the proper tone.

Regardless of the name, tone-access works like this: When you transmit to a repeater, you must add a low frequency tone between 69 and 255 Hz to your voice. (You can find a list of these tones in your radio's operating manual.) When the repeater receives your transmission, it checks your voice for the correct tone. If it detects the correct tone, the repeater forwards your voice to the repeater output. This system prevents interfering signals from activating the repeater transmitter — these signals won't carry the correct tone signal and thus are excluded from transmission. For a more detailed discussion of tone-access, see the *ARRL Handbook* or the December 1996 *QST* article, "Decoding the Secrets of CTCSS," by Ken Collier KO6UX.

Many recent radio models have a tone decoder function that detects which repeater is using a given tone. However, if your radio doesn't have this function, and you don't know the correct tone, you can't use the repeater. How can you find out what the proper tone is? Check with a repeater directory, which lists the tone and other vital statistics about the repeater. When you determine the correct tone — either via the tone decoder function or the repeater directory, you can then program your radio to send the correct tone and activate the repeater.

DCS (Digital Coded Squelch) is, like tone-access, another method for reducing interference. It allows you to hear audio from only selected individuals. DCS consists of a continuous sequence of low-frequency tones that accompanies the transmitted voice. If your receiver is set to the same code sequence, it passes the audio to the radio's speaker. If the transmission uses a different code, your radio remains silent. Most people use DCS to keep from having to listen to all of the chatter on a repeater, only hearing the audio of others using the same DCS code.

Be aware that not all repeaters pass the tone-access or DCS tones through to the transmitter and may filter them out.

Using simplex

When one station calls another without the aid of a repeater, both stations listen and talk on the same frequency, just as contacts are made on the HF bands, which is called *simplex* operation. Hams use simplex when they're just making a local contact over a few miles and don't need to use a repeater. Interspersed with the repeater frequency bands in Figure 9-1 are small sets of channels designated for simplex operation.

Having a common simplex channel is a good way for a local group of hams to keep in touch. Simplex frequencies are usually less busy than repeater frequencies and have a smaller coverage area, which makes them useful as local or town intercoms. Clubs and informal groups often decide to keep their radios tuned to a certain simplex frequency just for this purpose. If they're not having a meeting or conducting some other business, feel free to make a short call *("NØAX this is W7FMI")* and make a friend.

On bands with a lot of space, such as the VHF and UHF bands, making contacts outside of the repeater channels is easier if you know approximately where the other hams are. That's the purpose of *calling frequencies* — to get contacts started. You hear hams call *CQ* (the general "come in, anybody" call) on a calling frequency and, after establishing a contact, move to a nearby frequency to complete it. For example, if I call CQ on 52.525 MHz and N6TR answers me, I say, *"N6TR from NØAX, hi Larry, let's move to five-four, OK?"* This transaction means I receive N6TR and am tuning to 52.54 MHz, a secondary simplex frequency.

Making a couple of complete contacts on calling frequencies is okay if the band isn't busy, but otherwise move to a nearby frequency.

A national FM simplex calling frequency is set aside on each band just for general "Anybody want to chat?" calls. These frequencies are 52.525, 146.52, 223.50, 446.00, and 1294.5 MHz. When driving long distances, I often check these channels to meet up with other travelers on the highways. Visitors to your area often tune to these frequencies, so monitoring them is a good way to make travelers feel welcome or give directions.

If you are traveling and want to make a contact on the simplex calling frequencies, the best way to do so is to make a transmission similar to this one:

> *"This is NØAX November Zero Alpha X-ray mobile on Interstate 90. Anybody around?"*

Repeat this transmission a couple of times, spaced a few seconds apart. If you are moving, try making a call once every five minutes or so. Be sure to tell listeners you're traveling and your approximate direction. If you are operating away from your home area, add *"portable"* or *"portable (call district number)"* to your call as an additional bit of information.

Because simplex communications don't take advantage of a repeater's lofty position and powerful signal, you may have to listen harder than usual on these frequencies. Keep your squelch setting just above the noise level. When you're making a call, you may want to open the squelch completely, so you can hear a weak station responding.

Solid simplex communications usually require more power and better antennas than a typical hand-held radio, at least on one end of the contact. To get better results on simplex with just a few watts, try using a mobile antenna, a full-size ground plane, or a small beam. I discuss these antennas in Chapter 12.

Setting up your radio

After you figure out offsets, tones, and repeater frequencies, take a few minutes and check out your radio's operating manual. To make repeater (or simplex) contacts, you need to know how to do each of the following:

- **Set the radio's receive frequency and transmit offset.** Know how to switch to simplex (no offset) or to listen on the repeater's input frequency.

- **Turn sub-audible tones on or off and change the tone.** If your radio has the capability, know how to determine what tone frequency is being used.

✔ **Turn the Digital Squelch System on or off and change the codes.**

✔ **Store the radio settings in a memory channel and access different memory channels.**

Contacts, FM style

Operating and making contacts on FM is sufficiently different from the style of SSB or CW that it can confuse new licensees. Because VHF and UHF FM voice contacts are usually local or regional, they tend to be for personal utility, rather than to make random acquaintances. Most hams have a few favorite repeaters or simplex frequencies that they use as a sort of regional intercom. They turn on a radio in the shack at home or in the car and monitor a channel or two to keep an ear out for family or friends. Even though a number of people may be monitoring a repeater, they mostly just listen unless someone calls them specifically or they hear a request for information or help.

This style can be a little off-putting to hams new to FM and can even seem unfriendly at times. Rest assured that the hams are not being unfriendly, they're just not in "meet and greet" mode on the repeater or a favorite simplex channel. Imagine the difference between meeting someone at a party versus at a grocery store. At the party, everyone makes new acquaintances and has conversations. At the store, people aren't there as a social exercise and may even seem a little brusque. Keeping this idea in mind, the following sections cover some suggestions to help get some experience with the FM style.

Joining the group

The best way to become acquainted with a group is to participate in its activities. Nets are a very common group use of FM repeaters.

The most common nets on FM are for emergency services groups, weather and traffic (the automobile kind), and equipment swap meets. Technical assistance nets exist as well as nets that handle radiograms into and out of your area. (I discuss traffic handling in Chapter 10.) Use the ARRL net search Web page at www.arrl.org/FandES/field/nets/client/netsearch.html to find local nets on the VHF and UHF bands. Your Elmer may be able to help you out with times and frequencies, as well.

Almost all nets call for visitors, generally at the end of the net session, and that's your chance. When you check in (all you have to do is give your call sign and maybe your name), ask for an after-net contact with the net control station or a station that had something of interest to you. After-net contacts are initiated on the net frequency after the net is completed. Sometimes they are held on the net frequency and other times the stations establish contact there and move to a different frequency.

During the after-net QSO, you can introduce yourself and ask for help in finding other nets in the area. If you have specific interests, ask if the station knows of other nets on similar topics. Hopefully, you get a referral and maybe even a couple of call signs to contact for information.

By checking into a few nets, your call sign starts to become familiar and you have a new set of friends. If you can contribute to a weather or traffic net, or accept a radiogram destined for your town or neighborhood, by all means do so. Contributing your time and talents helps you become part of the on-the-air community in no time.

Seizing the opportunity

As you monitor the different channels, which repeaters encourage conversations and which don't becomes apparent. If you can identify a repeater that is ragchew-friendly, you have a fairly easy time making a few casual contacts. Listen for a station accessing the repeater, which sounds something like: *"NØAX monitoring"* or *"NØAX for a contact."* When you hear that, respond with a quick one-by-two call using phonetics, such as *"NØAX this is KD7DQO, Kilo Delta Seven Delta Queen Ocean, over."* By convention, calling CQ is reserved for SSB and CW operations where you're not sure who is out there and signals are generally weaker. To fit in with FM's intercom style, just make short transmissions.

Continue your contact as you do other contacts (see Chapter 8), but keeping your transmissions short is important. Repeaters have a time-out function that disables the transmitter if the input is busy for more than just a few (typically three to five) minutes. This time-out function prevents overheating the transmitter and also keeps a long-winded speaker from locking out other users. When you're transmitting, if you let up on the mike switch and no repeater signal comes back, you may have timed-out the repeater. Just let the repeater rest for 10 or 15 seconds and the receiver re-enables and the timer resets. No need to be embarrassed (unless you keep doing it) because all hams have done it more than once.

Open and closed

If you purchase a repeater directory, you may see some repeaters are marked as closed, and are not open to the ham radio public. Some repeaters are closed because they are dedicated to a specific purpose, such as emergency communications. Some repeaters are intended to be used by only a supporting group. Rest assured that you can use a closed repeater in case of emergency, but respect the wishes of its ownership and don't use it for casual operation. If you aren't sure whether the repeater is closed or not, transmit to it and ask, *"This is NØAX, is a control operator on frequency?"* If you get a response, that's the person to ask.

Repeater features

You can find an amazing set of features out there in repeater land. Many repeaters have voice synthesizers that identify the repeater and announce the time and temperature. Hams using the repeater can activate and deactivate some functions, such as autopatch and automated announcements of time or temperature, by using Touch-Tone tones from the keypads commonly found on microphones. Repeaters are linked together to provide wide coverage even across bands.

In the next few sections, I only touch on the basics of repeater operation and features. Repeaters are widely used all over the world and a wealth of other information is available for you to consult as you grow more experienced and curious about repeater and FM operations. Check the following Web sites for more information:

- ✔ **AC6V Web site** (`www.ac6v.com/repeaters.htm`): This site offers comprehensive links on many different ham radio topics.

- ✔ **eHam.net** (`www.eham.net/newham`): This ham radio portal includes numerous areas of interest to hams, including a handy New Ham page. Click the Basic Operating link for information about repeater operating.

- ✔ **ARRL Technical Information Services** (`www.arrl.org/tis/info/repeater.html`): This site has many public links and numerous in-depth articles for ARRL members.

- ✔ **K1HWU Ham Links** (`www.k1dwu.net/ham-links/repeaters.phtml`): This site offers a lot of links to repeater groups in the United States.

Autopatch

The best known feature of repeaters is called *autopatch*. This feature allows a repeater user to make a telephone call through the repeater! By accessing a repeater's autopatch function, a dial tone appears on the air. When you use the numeric keypad commonly found on your radio or microphone, the tones feed through the telephone system, which dials the number and connects you.

All autopatch calls occur over the air — they're not private.

Although you can do a lot of important things with autopatch — most hams use it to call 911 to report accidents, for example — you should use this function wisely. You must use special access codes to activate the feature. To get these codes, you may be required to join the club or group that maintains the repeater. Some limits are in place on what you are allowed to do via autopatch. For example, conducting business is forbidden as on any amateur frequency. (You can perform a limited amount of personal business, such as calling a store or — my favorite — ordering a pizza.) And most repeater systems place strict limits on the area codes that you can dial to prevent incurring long-distance charges.

With mobile phone service nearly ubiquitous today, autopatch is not the unique capability it once was. Even so, in certain conditions, like if your cellphone can't find a suitable service provider or if the mobile phone systems are overloaded or unreachable, autopatch may be able to get through.

If your radio has the autopatch feature and you obtain the tone sequences, follow these steps to make a phone call (the exact steps may vary from repeater to repeater):

1. **In an emergency, break in to an ongoing contact and ask the stations to stand by for an emergency autopatch.**

   If it is not an emergency, wait until the channel is clear.

2. **Tell the repeater you're switching to autopatch and without releasing the microphone button on the microphone, press the activation sequence of tones. Release the mike button.**

   For example, if your call sign is NØAX, you say, *"NØAX accessing autopatch,"* and press *82 or #11.

3. **The repeater acknowledges your sequence with some kind of tone, synthesized voice, or maybe just the telephone system dial tone.**

   If you don't hear an acknowledging transmission, go back to Step 2. You may have trouble if your signal is fluttery or noisy.

   You can also ask for another repeater user to activate the autopatch function for you.

4. **When you hear the dial tone, press the tone keys for the number you want to dial.**

   You may have to precede the telephone number with a dialing code, depending on the repeater's specific operating requirements.

5. **When someone answers your call, tell him or her you're using a radio.**

   For example, you can say, *"This is (your name) and I am using an amateur radio autopatch system."* This transmission lets him or her know that a radio link is involved so you both can't talk at once. Emergency services dispatch operators are generally familiar with autopatch, but pizza delivery operators may not be.

6. **Transmit your message.**

   Keep it short and appropriate for a public conversation.

7. **When you finish, key in the hang-up tones (if needed) while keeping the mike button pressed and tell the repeater you're disabling autopatch.**

   For example, say *"This is NØAX releasing autopatch."*

8. **The repeater acknowledges the release with an announcement similar to that at activation.**

9. **(Optional) You may let the other repeater users know that you are finished and can resume normal operation.**

Repeater networks

Some repeaters can be linked together in networks that provide coverage over wide areas. For example, the system of repeaters known as the Cactus Inter-tie allows users to communicate using their local repeater in an area that covers Los Angeles to the east and north as far as Texas and Utah, respectively. In my area, the Evergreen Inter-tie links repeaters throughout the Pacific Northwest and Canada. The Target Repeater System hooks up hams from Ohio through Virginia. Many more of these networks exist, as evidenced by the list at www.qsl.net/w9sar/repeaters.html. Repeaters may be linked at all times or only under the control of authorized operators.

To use these systems, you need to find out which local repeaters are "on the inter-tie," usually by browsing the Web site for the repeater system. If you decide to become a member of the group that maintains the network, you get access to the control tone sequences that activate the network connections. You can usually find the necessary information about joining a group on its Web site; or you can listen for a group member controlling the repeater and give him or her a call. Membership requirements and expectations vary, so be sure you can meet them before signing up. Setting up a repeater link across several states from a hand-held radio is pretty neat!

Linking repeaters with the Internet

The latest advance in repeater linking is a novel hybrid system that marries the VHF/UHF repeater to the Internet. Several systems allow repeater users in one spot to tunnel through the Internet and pop up somewhere very far away, indeed! IRLP, iLINK, and Wires II are all examples of such systems. I use IRLP to illustrate the concept.

IRLP uses Voice-Over-IP (VOIP) technology to connect repeaters all over the world. This technology allows hams with hand-held portables to chat with more than 20 other countries around the world in more than 30 countries as of late 2003. You can even find IRLP on Ross Island in Antarctica! Figure 9-4 shows the basic IRLP system.

A *node* is a regular FM repeater with an Internet link for relaying digitized voice. A user or control operator can direct an IRLP node to connect to any other IRLP node. When the node-to-node connection is made, the audio on the two repeaters is exchanged, just as if both users were talking on the same repeater. Having contacts between Europe and New Zealand with hams on both ends using hand-held radios putting out just a watt or two is common!

You can also connect several nodes together using an IRLP reflector. The reflector exchanges digitized audio data from any node in real time with several other nodes. Even a user without a radio can join in by logging on to an IRLP reflector or node. All users that cause radio transmissions to be made have to be licensed though.

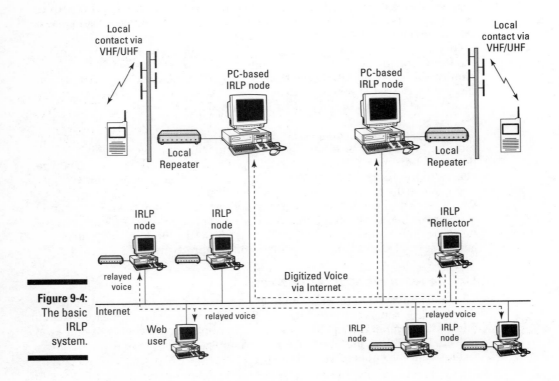

Figure 9-4: The basic IRLP system.

Using the IRLP system is very much like using an autopatch system. You don't need anything more than your radio, the IRLP system control tone sequences for your repeater, and a list of the four-digit node on-codes that form the IRLP address of active IRLP repeaters. To connect to another IRLP-enabled repeater, follow these steps:

1. **Enter an IRLP access code.**

   The access code sets up the repeater to accept an IRLP on-code. This process is just like activating autopatch.

2. **When the repeater indicates that the IRLP system is ready, enter the tones that send the on-code of the repeater that you wish to connect to.**

   Enter the tone is just like entering a telephone number into an autopatch system. You can find a list of the available IRLP nodes and their on-codes at status.irlp.net.

3. **After a short delay, the node you connected to identifies itself in plain voice with a call sign and location and you are connected!**

   Any transmissions you make transmit on the remote node and you hear all the audio from the other node. If the other node is busy with another IRLP connection, you hear a message to that effect. Try another node or come back later.

4. **When you finish, use the IRLP control codes to disconnect, as with an autopatch.**

You can get more complete information on the IRLP system through several Web resources. An introductory article at www.eham.net/newham/irlp gives a good overview of the system. If you're really interested in the complete technical details and maybe even putting together your own IRLP node, the IRLP home site at www.irlp.net has complete details.

For more information on other Internet-linking of voice systems, check out the following URLs:

- **iLINK:** www.geocities.com/gj7jhf/ilinking.html
- **Wires II:** www.vxstd.com/en/wiresinfo-en
- **Echolink:** www.echolink.org

Chewing the Rag

I first mentioned ragchewing in Chapter 8 as a likely way that you make your first contacts. In this section, I go into a little deeper detail about the ham etiquette of the ragchew.

Knowing where to chew

You know enough about ham radio that you can't just spin the dial and start bellowing into a microphone. Although ragchewing isn't listed on any band plan, you can find ragchewers on certain parts of every band.

On the HF frequencies below 30 MHz, the bands all have a very similar structure: CW (Morse code) and data modes occupy roughly the lower third and voice modes occupy the upper two-thirds. This structure is true even on the WARC bands of 17 and 12 meters where no formal division between the modes is made.

Figure 9-5 shows how the general operating styles are organized on a typical HF band. You find the ragchewers mixed in with the DX contacts on the lower end of the band. Sometimes, if a super-rare station comes on the air, the sheer numbers of DXers calling crowd a sustained contact (a *pileup*), so the ragchewers move up the band. The ragchewers space themselves out around the nets, round tables, calling frequencies, and data signals, taking advantage of ham radio's unique frequency agility to find an empty spot on the band.

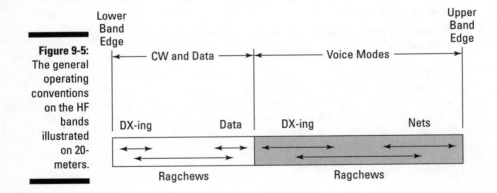

Figure 9-5: The general operating conventions on the HF bands illustrated on 20-meters.

What's a WARC?

WARC refers to the international World Administrative Radio Conference of 1979. At that conference, amateurs worldwide were granted access to three new bands: 30, 17, and 12 meters. These new bands were immediately nicknamed the WARC bands to distinguish them from the older ham bands at 160, 80, 40, 20, 15, and 10 meters.

DX-ing takes place at the lower end of the band for two primary reasons:

- **Long-distance contacts are generally more difficult to make than regional or local contacts.** The signals are weaker and DX enthusiasts find each other easier if they tend to operate in one area of the band.

- **After the bands were divided into different sub-bands for the different license classes, fewer higher-class operators were in the lowest segments of the bands.** The rarer DX stations took advantage of the lesser population of U.S. hams in those lower band segments to reduce the crowd pursuing them on the air.

Because DX-ing tends to attract a crowd, this concept is somewhat incompatible with the more ordered operating style of nets. Therefore, those types of operations would gather at the "other end" of the bands. With this structure, groups can engage in their preferred style of operating without interference. Data and Morse code styles tend to be incompatible and so the data signals, such as radioteletype (RTTY) or PSK, also stay towards the higher portion of the band. Not many nets use Morse code and data, so data signals don't interfere too much.

You may think that ragchewers are buffeted from all sides, but that's not really the case. A lot of ragchew contacts take place all the time and so they tend to occupy just about any spare bit of band. To be sure, in the case of emergencies (when a lot of nets are active), major expeditions to rare places, or on weekends when big contests are running, the bands may seem too full to get a word in edgewise.

The FCC can declare a communications emergency and designate certain frequencies for emergency traffic and other communications. Keeping those frequencies clear is every amateur's responsibility. The ARRL broadcasts special bulletins over the air on W1AW, by e-mail, and on its Web site if the FCC does make such a declaration. The restrictions are in place until the FCC lifts them.

On the VHF and UHF bands, you can usually find ragchewing in the repeater sections of the bands, although wide open spaces for a conversation are available in the so-called "weak signal" portions of the bands. Table 9-2 lists the calling frequencies and portions of the VHF/UHF bands. The operating style in this portion of the bands is similar to HF as far as calling CQ and making random contacts, but the bands are less crowded.

Table 9-2	VHF/UHF Morse Code (CW) & SSB Calling Frequencies	
Band	**Frequencies (in MHz)**	**Use**
6 Meters	50.0–50.3	CW & SSB
	50.070, 50.090	CW Calling Frequencies
	50.125 and 50.200	SSB Calling Frequency, use Upper Sideband (USB)
2 Meters	144.0–144.3	CW & SSB
	144.050	CW Calling Frequency
	144.200	SSB Calling Frequency, use USB
222 MHz (1-1/4 Meters)	222.0–222.15	CW & SSB
	222.100	CW & SSB Calling Frequency, use USB
440 MHz (70 cm)	432.07–433.0	CW & SSB
	432.100	CW & SSB Calling Frequency, use USB

The lower portion of the VHF/UHF bands is referred to as a weak signal although that's really a misnomer. The reason for the name is that contacts using CW and SSB can be made with considerably weaker signals than on FM. Most of the CW and SSB signals you hear are quite sufficiently strong for excellent contacts, thank you!

Knowing when to chew

Whether you're on HF or VHF/UHF, you find ragchewing has its good times and poor times. When calling CQ (signifying that you want to talk to any station), you can let it be known that you're looking for an extended contact in a number of ways. You also hear numerous clues that a ragchew may not be what another station has in mind.

I keep coming back to this point, but listening is the best way to learn operating procedures in ham radio. The most important part of any amateur's station is between the ears! If you want to call CQ successfully, then spend some time listening to more experienced hams do it.

The good times

Assuming that you're tuning an *open* band (signals are coming in from various points), when is a good time to ragchew? First, consider the social aspects of your contact timing. Weekdays are generally pretty good days to ragchew, especially the daylight hours when hams that have day jobs are at work. ***Remember:*** It may be a different time of day at the other end of the contact, particularly if the other station is DX.

You may want to revisit Chapter 8 when considering what band is best to use. If you like to talk regionally, you can always use a repeater or one of the lower HF bands. To talk coast-to-coast, one of the higher bands is your best bet. The better your antenna system, the more options you have.

Lots of hams do their operating on weekends, but that's also when special events and contests are held. Be prepared for a full band on Saturday and Sunday every weekend of the year. The silver lining to the cloud is that plenty of hams are on the air for you to contact! If you know that one mode or band is host to some major event, you can almost always find a quiet spot on the other mode or other bands. The WARC bands never have contests and are usually wide open for ragchews and casual operating.

Not-so-good times

Because a good ragchew lasts for a long time, pick a time and band that offer stable conditions. Propagation changes rapidly around sunrise and sunset. Local noon can be difficult on the higher bands. Don't be afraid to try any old time; you may surprise yourself and you'll learn about propagation from the best teacher — experience!

Because weekends are busy times, you should check the contest and special-event calendars (see Chapter 10). A little advance warning helps keep you from being surprised when you get on the air and allows you to be flexible in your operating.

When the bands seem frustratingly full, here are some helpful strategies that keep you doing your thing:

- **The majority of nets and all contests are run on the traditional bands.** The WARC bands almost always have sufficient space for a QSO.

- **Only a few contests sponsor activity on both phone and CW.** No contests have activity on phone, CW, and data. You can change modes and enjoy a nice ragchew.

- **Be sure you know how to operate your receiver.** Cut back on the RF sensitivity, use your narrower filter settings, know how to use controls such as the IF Shift or Passband Tuning controls, and generally be a sharp operator. You can remove much interference and noise just by using all the adjustments your receiver provides.

- **Always have a backup plan!** There's no guarantee that any particular frequency is clear on any given day. Hams have "frequency freedom" second to none, so use that big knob on the front of your radio!

Identifying a ragchewer

If you're in the mood for a ragchew and you're tuning the bands, how can you tell if a station wants to ragchew? The easiest way is finding an ongoing ragchew and joining it. You can break in (see Chapter 8) or wait until one station is signing off and then call the remaining station.

Otherwise, the key is in the CQ. Cues that the other station is looking to ragchew include a relaxed tone of voice and an easy tempo of speaking. Look for a station that has a solid signal — not necessarily a needle-pinning strong station, but one that is easy to copy and has a steady signal strength. The best ragchews are contacts that last long enough for you to get past the opening pleasantries, so find a signal that you think can hold up.

One cue that the station is not looking for a ragchew is a *targeted call.* For example, you may hear, *"CQ New York, CQ New York from W7VMI."* W7VMI likely has some kind of errand or message and is interested in getting the job done. Perhaps the station is calling, *"CQ DX"* or *"CQ Mobiles,"* in which case, if you're not one of the target populations, keep on tuning.

Another not-a-ragchew cue is a hurried call or a call that has lots of stations responding. This station may be in a rare spot, in a contest, or at a special event. Keep on tuning if you're really looking for a ragchew.

Evaluate on-the-air technique as you tune across the bands. Consider what you like and dislike about the various styles. Take what you like and try to make it better — that's the amateur way.

Sharing a ragchew

Hams come from all walks of life and have all kinds of personalities, of course, so you come across the garrulous, for whom a ragchew that doesn't last an hour is too short, and the mike-shy ham that considers more than a signal report to be a ragchew. Relax and enjoy the different people you meet!

Round tables are also great ways to have a ragchew. These are contacts between three or more hams on a single frequency. Imagine getting together with your friends for lunch. If only one of you could talk at a time, that's a round table! These aren't formal, like nets, and generally just go "around the circle" with each station talking in turn. Stations can sign off and join in at any time.

Pounding Brass — Morse Code

CW is a lot easier to learn and copy if you are equipped to listen to it properly. For starters, using headphones *(cans)* really helps because they block out distracting noise. When you are copying code, your brain evaluates every little bit of sound your ears receive, so make its job easier by limiting the non-code sounds. **Warning:** Settling on one preferred *pitch* for the tones is natural, but over long periods of time, you can "wear out" your ear at this pitch. Keep the volume down and try different pitches so you don't fatigue your hearing.

When you have a comfortable audio environment, be sure your radio is set up properly, too. Most radios come with a receiving filter intended for use with voice signals. Typically, a voice filter is 2.4 kHz wide (meaning that it passes a portion of radio spectrum or audio that spans 2.4 kHz) to pass the human voice clearly. CW doesn't need all that bandwidth. A filter 500 Hz wide is a better choice and you can purchase one as an accessory whether your radio is new or used. The narrower filter removes nearby signals and noise that interfere with the desired signal. In fact, four or five code signals can happily coexist in the bandwidth occupied by a single voice signal!

Narrower is not necessarily better below 400–500 Hz. A very narrow filter, such as 250 Hz models, may allow you to slice your radio's view of the spectrum very thin, but the tradeoff is an un-natural ringy sound to the signal. You are less able to hear what's going on around your frequency. These extra-narrow filters are useful when interference or noise is severe, but use a wider filter for regular use.

Be sure to read through your radio's operating manual sections on CW operation. Find out how to use all of the filter adjustment controls, such as the IF Shift and Passband Tuning controls. Most CW operators like to set the

AGC control to the FAST setting, so that the radio receiver recovers rapidly. Being able to get the most out of your receiver is just as important on CW as when using voice.

If your radio has the ability to switch rapidly between transmit and receive, then the *full-breakin* mode (or *QSK*) is something you'll want to try. In this mode, you can hear what's happening on the band between the dits and dahs. You have to turn off VOX operation that holds the radio in transmit.

Copying the code

To get really comfortable with CW, you need to copy in your head. Watching a good operator having a conversation without writing down a word is an eye-opener. How do they do that? The answer is practice.

As your code speed increases during the learning process, you gradually achieve the ability to process whole groups of characters as one group of sound. Copying in your head just takes that to another level. To get there, spend some time just listening to code on the air without writing anything down. Without the need to respond to the sender, you can relax and not get all tensed up trying not to miss a character. Soon, you can hold more and more of the contact in your head without diminishing your copying ability.

When you try it for real, use a piece of paper; not to write down all the characters, but to jot down topics and information for your part of the next transmission. Resist the temptation to write each letter on the paper. Soon you're ready for the next step, *copying behind.*

Read some more mail on the bands, trying to relax as much as possible without staying right up with each character. Don't force the meaning and let your brain give it to you when it's ready. Gradually the meaning pops into your head farther and farther behind the characters as they're actually received. What's happening is that your brain is doing its own form of error correcting, making sure that what you copy makes sense and taking cues from previous words and characters to fill in any blanks.

Good copying ability sneaks up on you over time. When you really hit a groove, you're barely conscious of the copying process at all and, *voila!,* CW is your second language.

Sending Morse

You may think that sending ability automatically follows receiving ability. To some extent, that's true, but after listening to other operators on the air, you'll find a wide range of sending ability. Having a good, smooth sending style, or *fist,* is not hard.

First decide what type of device you want to use to send code. The basic telegraph key or *straight key* seen in Figure 9-6 is heard on the bands every day, but sending good code at high speed with one is challenging. The straight key tops out somewhere between 20 and 30 words per minute (wpm). At these speeds, sending becomes a full-body experience and you really have to be skilled to make it sound good. The straight key has a few variants — the sideswiper, for example, in which a keying lever is moved back and forth. Still, manual keying on a sideswiper tops out around 30 wpm.

You can find several better options. Before the advent of inexpensive electronics, fast code was sent with an *automatic key,* now known as a *bug,* and shown in the middle of Figure 9-6. They're called bugs because the largest manufacturer, Vibroplex (www.vibroplex.com), uses the lightning bug as its symbol. A cautionary note: Bugs are rarely heard today, which makes their rhythm unusual and hard to copy, especially in the fist of an unskilled operator. You still make the dashes manually by pressing the keying paddle to the left, but dots are now made by a vibrating lever attached to a pair of contacts. A weight sliding along the lever determines the speed of the dots. These vibrating-lever contraptions have a distinctive sound all their own, a syncopation known as *swing.*

As vacuum tubes miniaturized and transistors came on the scene in the mid-1960s, electronic *keyers* were created. These devices generate dots and dashes electronically, when a dot contact or dash contact is closed for as long as the contact stays closed. The simplest electronic keyers only make strings of dits and dahs and the operator puts them together with the right timing. More sophisticated keyers make sure that the spacing between dits and dahs is correct and can even send alternating dits and dahs if both contacts are closed at the same time. These keyers are called *iambic* keyers because of the didahdidahdidah pattern they make. Because of the way Morse characters are constructed, a skilled keyer makes fewer hand movements using an iambic keyer. The devices used to send code with keyers are called *paddles* (shown in Figure 9-6) after the flat ovals touched by the operator. A good operator can send well over 60 wpm with an electronic keyer and comfortable paddle!

Collectors of Morse code equipment extend far beyond the ham radio community. Railroad and telegraph aficionados have terrific collections of old keys, bugs, and paddles. For an entry into the world of antique code, start at the Sparks Telegraph Key Review (www.zianet.com/sparks). Morse Express Books (www.mtechnologies.com/books) publishes excellent books about keys and related equipment.

Combination Paddle & Key

Straight Key

Semi-automatic key or "Bug"

Figure 9-6:
My paddle-
key combo,
a bug,
and the
venerable
straight key.

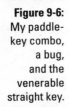

Straight Key Night is a fun event that brings out old and new code equipment (and operators) around the world . Every year on New Year's Eve, hams break out their straight keys and bugs and return to the airwaves for a few old-time QSOs before heading off to the evening's frivolities. Auld lang syne and all that, you know! An award is given for Best Fist, too. Give it a try this year!

Code by computer

A number of programs use a computer's sound card to extract characters from the code. Some of them can even copy more than one code stream at a time! (Computers can go fast, but in the presence of noise or interference, a human is still much better at copying.) Keying is done by a transistor interface to the computer's serial or parallel port. Personally, I prefer to send and copy the code myself, but can't begin to approach the speed at which computers can handle Samuel's invention. Review the list of Morse programs at www.ac6v.com/morseprograms.htm to find a program to try out.

Making and responding to Morse code calls

On the HF bands, you can always find code on the lower frequencies of the band, even on bands not divided between code and non-code modes. The faster operators tend to be at the very bottom of the bands, with average code speed slowly dropping as you tune higher.

As you start, find an operator sending code at a speed you feel comfortable receiving. Even though no new Novice licenses are being granted, a number of slow-speed code operators are in the Novice sub-bands 3.675 to 3.725, 7.100 to 7.150, 21.100 to 21.200, and 28.100 to 28.300 MHz. Medium-speed QSOs are the norm elsewhere; even mixed in with high-speed operators down low in the band.

When sending a Morse code CQ (a general way to solicit a contact), don't send faster than you can receive. Having to ask the responding station to *QRS* (slow down) because you hustled through your CQ is embarrassing.

The best CQ is one that's long enough to attract the attention of a station tuning by, but not so long that that station loses interest and tunes away again. I have good luck with 3-by-2-by-3 CQs (CQ CQ CQ de NØAX NØAX, repeated three times) on average. If the band is very quiet, you may want to send longer and a busy band may only require a 2-by-2-by-2 CQ. You just have to try different styles until you get a feel for what works.

If your radio has the ability to listen between the dits and dahs (your operating manual calls this the Break-in, or QSK, feature), use it to listen for a station sending "dits" on your frequency. That means, "I hear you, so stop CQing and let me call!" You can then finish up with *"DE [your call] K"* and the other station can call right away.

Making Morse code contacts (CW)

Making code contacts, or CW, is a lot like making voice contacts in terms of structure. Hams are hams, after all. What's different about the Morse code contact is the heavy use of abbreviations, shorthand, and prosigns (two-letter combinations used to control the flow of a contact) to cut down the number of characters you send. Find a complete list of CW abbreviations and prosigns at ac6v.com/morseaids.htm.

After you begin a contact and exchange call signs, including your call sign every time you turn the transmission over to the other station is not necessary, but you must include it once every ten minutes as required by FCC rules. Send your information and end with the BK prosign to signal the other station that he or she can go ahead. This method is much more efficient than sending call signs every time.

At the conclusion of a Morse code contact, after all the 73s (Best Regards) and CULs (See You Later), be sure to close with the appropriate prosign: SK for end of contact or CL if you're going off the air. You may also hear the other station send "shave and haircut" (dit-dididit-dit) and you are expected to respond with "two bits" (dit dit). These rhythms are deeply ingrained in ham radio and are even heard in spoken conversations between hams. I wrap up many a chat with "diddly bump-de-bump," the rhythm of SK, or just a "dit dit" meaning, "See ya!" Yeah, it's a little goofy, but have fun!

Morse code (CW) clubs

After you master Morse code, it stays with you forever and you may find it the most pleasurable way of making contacts. A number of groups for Morse code buffs are around the world. FISTS (www.fists.org), the International Morse Preservation Society, has chapters around the world and is the largest organization dedicated to Morse code. The British club First Class CW Operator's Club (www.firstclasscw.org.uk), or FOC, is quite exclusive with only 500 active members at any given time, but is quite active. Enter **CW Club** into an Internet search engine to find other smaller CW clubs.

Receiving Messages Afloat and Remote

Once upon a time, any kind of messaging capability was tied to specific phone numbers, bulletin boards, and servers. If you weren't in range of your home bulletin board or service provider, then you were pretty much disconnected. Ham radio has left those days behind, just like wireless networks have freed up the computer user.

Hams can use a computer and a radio to connect directly to gateway stations around the world by terrestrial links on the HF bands, or by VHF and UHF to the amateur satellites. The gateway stations transfer e-mail messages between ham radio and the Internet. The satellites transfer messages to and from the

Internet via a ground control station. Either way, a ham far from home can use ham radio to send and receive e-mail.

The best-known HF message system is Winlink 2000 (www.winlink.org), which enables anyone with a Windows-based computer to send and receive e-mail over ham radio links using the PACTOR or PACTOR II digital modes (I cover digital modes in Chapter 11). A worldwide network of stations, shown in Figure 9-7, operates on the 80, 40, 30, and 20 meter bands as well as VHF packet. To find the frequencies for these stations, click the Winlink Stations link on the Winlink home page. This extensive and growing network covers much of Earth for most propagation conditions. These stations are linked via the Internet, creating a global home for Winlink users.

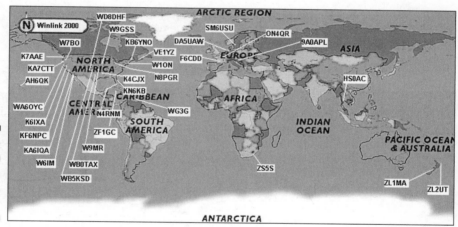

Figure 9-7: Winlink 2000 digital data network stations.

To use the Winlink 2000 system, you first must register as a user on the Winlink network so that the system recognizes you when you connect. As a recognized user, your messages are available from anywhere on Earth, via whichever Winlink station you use to connect. You must also download and install a Winlink-compatible e-mail program, such as AirMail, which is available via the Winlink Web site.

Along with a computer that runs the e-mail software, you also need a sound card and software to send and receive PACTOR data or an external communications processor that can run PACTOR or PACTOR II. Your HF or VHF radio connects to the Winlink station.

If you are using AirMail, the connection process is straightforward. Follow these steps to connect:

1. **Open the AirMail program.**

2. **Click the Terminal icon.**

   A menu of WinLink stations and frequencies pops up.

3. **Select a station and frequency appropriate for your location, time of day, and equipment.**

   The computer and radio attempts to connect to the WinLink server and notifies you of success or failure. This process is very much like a modem connecting to a remote server for a dialup connection.

After you connect to the Winlink 2000 system, your messages are available anywhere. The data rate is limited due to the radio link, so don't try to send big files, but text messages go through just fine.

The usual regulations for ham radio messages apply, of course; you cannot encrypt messages, send business traffic and obscene content, or use radio links on behalf of third parties in countries where such use is prohibited. Also keep in mind that while the Winlink stations are connected via the Internet, your station is connected by a relatively slow digital data radio link. Avoid large files and messages. Nevertheless, Winlink 2000 is a tremendous service and a boon to those who travel off the beaten track.

Chapter 10

Operating with Intent

• •

In This Chapter

▶ Registering with an emergency organization

▶ Getting ready for and operating in emergencies

▶ Rendering public service

▶ Transmitting on nets

▶ Delivering messages

• •

*A*s your experience with ham radio grows, you'll find more and more practical uses for your communications skills. Your ham radio skills can also benefit others, which is where the *service* part of *Amateur Radio Service* comes in.

In return for the privileges that go with the license — access to a broad range of frequencies, protection from many forms of interference, maintenance of technical standards, and enforcement of operating rules — the Amateur Service gives back by providing emergency communications systems and trained operators. In between emergencies, hams also provide communications support at public functions and sporting events, keep an eye on the weather, and perform training exercises. And hams have an extensive message handling network that runs every day of the year for both emergency and ordinary messages.

These services are important to you for two reasons: You can use them and you can provide them. In this chapter, you find out what those services are and how to get started with the groups that provide them. This chapter is written primarily for American and Canadian hams and describes the U.S. emergency communications organizations. Elsewhere around the world, you can find similar organizations to varying degrees. Contact your national amateur radio society for information about them.

Joining an Emergency Organization

As I discuss at the beginning of Part II, the very first item in the list that describes the basis and purpose of the Amateur Service is emergency communications. As I write this (October 2003), other hams are putting their skills to work helping fight the Southern California wildfires. Over the course of this year, hams have been called to action as a consequence of severe weather, fires, earthquakes, and even to assist with debris collection from the breakup of the Space Shuttle, *Columbia.* You never know when an emergency will arise, so start preparing yourself as soon as you're licensed.

Known in the radio biz as *emcomm,* emergency communications is loosely defined as any communication with the purpose of reducing an immediate threat of injury or property damage. This definition covers everything from reporting car accidents to supporting large-scale disaster relief. In this section, I introduce you to the elements of emcomm and show you where to find the necessary information to get started.

Regardless of whether your interest in emcomm is to support you and your family or to participate in organized emcomm, you need to know how amateurs are organized. Otherwise, how do you know where to tune or how to interact with them? That's where the ARRL's Field Organization comes in. While other local and regional amateur emcomm organizations are certainly active, the Field Organization's Amateur Radio Emergency Service (ARES) is the largest nationwide ham radio emcomm organization. Like the rest of the ARRL membership, ARES is organized by individual ARRL sections that may be an entire state or as small as a few counties, depending on population. For complete information about the ARES, find the *Public Service Communications Manual* online at www.arrl.org/FandES/field/pscm.

The ARES and Radio Amateur Civil Emergency Service (RACES), organized and managed by the Federal Emergency Management Agency (FEMA), are national emergency communications organizations that provide communications assistance to public and private agencies during a civil emergency or disaster. They're open to any amateur and welcome your participation.

If you want to help administer and manage emcomm activities in your ARRL section after you have some ham radio experience, consider applying for an ARRL Field Organization appointment. Volunteers fulfill the following positions:

✔ **Assistant Section Manager (ASM):** Section managers are appointed, but you can always assist him or her. Tasks vary according to the activities of the section, but collecting and analyzing volunteer reports or working with and checking into local and regional nets are typical duties. Should a special task arise, you may be asked to take it on behalf of the Section Manager.

✔ **Official Emergency Stations (OES):** Perform specific actions as required by your local Emergency Coordinators. OES appointments are for stations committed to emcomm work and provide the opportunity to tackle detailed projects in operations, administration, or logistics.

✔ **Public Information Officer (PIO):** You can establish relationships with local and regional media in order to publicize ham radio, particularly the public service and emcomm performed on behalf of the public. PIOs also help establish good relationships with community leaders and organizations.

✔ **Official Observer (OO):** OOs help other hams avoid receiving an FCC notice of rule violation because of operating or technical irregularities. They also keep an ear out for unlicensed intruders or spurious transmissions from other services.

✔ **Technical Specialist (TS):** If you have expertise in a specific area or if you are generally skilled in some aspect of radio operations, you can be a Technical Specialist. The TS serves as a consultant to local and regional hams, as well as to the ARRL.

Check out the other interesting appointments at both the section and division level at www.arrl.org/FandES/field/org.

ARES and RACES

Why have two different organizations? Aren't they doing the same thing? Yes and no. The different levels of emergencies and disasters, with varying degrees of resource requirements, require different responses by government agencies. As a result, a single, one-size-fits-all amateur emergency organization is not enough to handle all emergencies.

The Amateur Radio Emergency Service (ARES) is managed by the ARRL Field Organization and works primarily with local public safety and non-governmental agencies, such as the Red Cross. Local ARES leadership determines how best to organize the volunteers and interact with the agencies they serve. Training is arranged by the ARES teams and local organizations.

You can register as an ARES volunteer by simply filling out an ARRL form (www.arrl.org/FandES/field/forms/fsd98.pdf) and mailing it to the ARRL. However, you also need to join a local ARES team to actually participate in training and exercises. The easiest way to find out about the ARES organization in your area is to contact your ARRL Division's Section Manager (SM) listed at www.arrl.org/FandES/field/org/smlist.html. You can also search the ARRL Net Web site at www.arrl.org/FandES/field/nets/client/netsearch.html and find ARES nets in your area. Check in to the net as a visitor and ask for information about ARES in your area.

Military Affiliate Radio Services (MARS)

A third organization that maintains an extensive emergency communications network of ham volunteers is Military Affiliate Radio Services (MARS), which provides an interface between the worldwide military communications systems and ham radio. MARS is sponsored by the Department of Defense, but each branch of the military has its own MARS program.

MARS members receive special licenses and call signs that allow them to operate on certain MARS frequencies just outside the ham bands. MARS provides technical and operations training, as well as preparation for emergency communications. MARS volunteers handle many personal messages from military personnel and provide *phone patches,* or connections between radios and the phone system, to their families while they are deployed away from home.

To be a MARS volunteer, you must be at least 18 years old, be a U.S. citizen, and hold a valid amateur license. For more information on each of the MARS programs, including information about how to apply for membership, see the following Web sites:

- ✔ **Army:** www.asc.army.mil/mars
- ✔ **Navy/Marines:** www.navymars.org
- ✔ **Air Force:** afmars.tripod.com/mars1. htm

The Radio Amateur Civil Emergency Service (RACES), founded as a Civil Defense support organization, is sponsored by the Federal Emergency Management Agency (FEMA) and is governed by special FCC rules. RACES is organized and managed by a local, county, or state civil defense agency responsible for disaster services and activated during civil emergencies by state or federal officials. RACES members are also required to be members of the local civil-preparedness group and receive training to support that group. More information on RACES is available at www.races.net.

To join RACES or find out more about RACES in your area, search your state government's Web site for Auxiliary Communications Service, also known as ACS. The local ACS coordinator is the person you need to contact about RACES membership. If you can't locate this information on your state's Web site, a different government agency may manage it. Enter your state's name along with **Auxiliary Communications Service** into a Web search engine and look for links that lead to emergency management sites.

I recommend that you start by participating in ARES. If you like being an ARES member, then dual membership in both ARES and RACES may be for you.

Preparing for an Emergency

Getting acquainted with emergency organizations is fine, but it's only a start. You need to take the necessary steps to prepare yourself so that when the time comes, you are ready to contribute. There are four parts to being prepared. You must know *who, where, what,* and *how.*

Knowing who

I discuss in the previous section the organizations that perform emergency communications. After you become familiar with the leadership in your ARRL section, you also need to get acquainted with the local team leaders and members. The call signs of the local clubs and stations operating from governmental Emergency Operations Centers (EOC) are valuable to have at your fingertips in times of emergencies.

The best way to make these call signs familiar to you (and yours to them) is to become a regular participant in nets and exercises. Checking into weekly nets takes little time and reinforces your awareness of who else in your area is participating. If you have the time, attending meetings and other functions such as EOC open houses or work parties also helps put a face with the call signs. Building personal relationships pays off when a real emergency comes along.

Knowing where

When an emergency occurs, you don't want to be left tuning around the bands trying to find out where emcomm is going on! Keep a detailed reference that lists the emergency net frequencies along with the names of the leaders in your area (I provide a chart for you to fill in on the Cheat Sheet). You may even wish to reduce this list with a photocopier and laminate it for a long-lasting reference the size of a credit card that you can carry in your wallet or purse.

Knowing what

You don't want to be running around trying to get your gear together in an emergency. If an emergency occurs and your equipment is not ready, you can be under tremendous pressure. In your haste, you either omit some crucial item or can't find it on the spur of the moment. I recommend a Go Kit (similar to a first aid kit) as an antidote to this adrenaline-induced confusion.

Before making up your Go Kit, consider what mission or missions you may be attempting. A personal checklist is a good starting point for your plans. A good generic checklist is available in the *ARES Field Resources Manual,* which you can download at www.arrl.org/FandES/field/aresman.pdf. Figure 10-1 shows an example of a portable Go Kit.

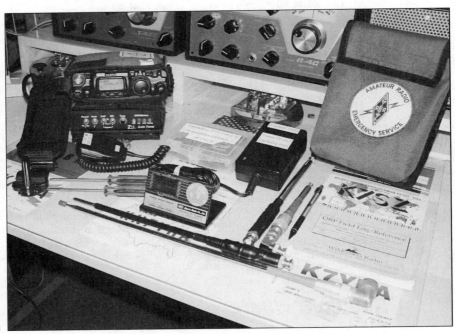

Figure 10-1:
A Go Kit.

A *Go Kit* is a group of items necessary to you in an emergency that you collect in advance of any such emergency and place in a handy carrying case. If an emergency situation actually arises, the Go Kit allows you to spend your time actually responding to the emergency instead of wracking your brain trying to get your things together on the spur of the moment. By preparing the kit in advance, you are less likely to forget important elements.

What should you put in your Go Kit? Well, of course, answers vary from ham to ham, but your Go Kit should at least contain the following essentials:

- ✔ **Food, particularly the kind not requiring refrigeration:** You never know when your next meal will arrive in an emergency. Remove the uncertainty and bring along your own.

- ✔ **Clothing appropriate to the emergency you're responding to:** If you get too cold, you want a jacket nearby; too hot and you can exchange your current clothing for something lighter. Preparation allows flexibility.

✔ **Radios and equipment:** Don't forget to bring everything you may need: radios, antennas, and power supplies. Make sure they're lightweight, flexible, and easy to set up.

✔ **References:** You need lists of operating frequencies as well as phone lists — a personal phone list and a list of emergency-related telephone numbers.

If you operate from home, no Go Kit is required, but you still need to prepare for emergencies — such as if the lights go out for an extended period or your main antenna goes down.

Your primary concern at home is emergency power. Most modern radios are not very battery friendly, drawing more than an amp even when just receiving. You need a generator to power them for any period of time. If you have a home generator, make sure it can be connected to power the AC circuits in your radio shack.

If you don't have a generator, you may be able to use another backup power source: Most radios with a DC power supply can run from an automobile. However, getting power from your car to your radio is not always easy. Decide which radios you want to operate from your car and investigate how you can power and connect an antenna to each of them.

Overall, the most important step is to simply consider and attempt to implement the answer to the question: "How would I get on the air if I'm unable to use my regular shack?" Just by thinking things through and making plans, you're on the road to being prepared in an emergency.

Knowing how

Knowing the procedures to take is the most important part of personal preparedness. No matter what your experience and background, the specifics of working with your emergency organizations must be personally learned. If they are not, you won't be prepared to contribute when you show up on the air from home or at a disaster site.

Do everyone a favor, including yourself, by spending a little time getting trained in the necessary procedure and techniques. Your local emcomm organization has plenty of training opportunities. You can check into the local NTS and emcomm nets for practice. Participating in public service activities, such as acting as a race course checkpoint in a fun run or as a parade coordinator, is awfully good practice and exercises your radio equipment, as well. By the way, you make good friends at these exercises who can teach you a lot!

The ARRL has also created a series of Emergency Communications training courses that you can take online. The courses require a tuition fee, but as of October 2003, because of government and private grants, students who complete the course successfully are also reimbursed for their tuition 100 percent. Check out the courses at www.arrl.org/cce.

After you start in emergency communications training, you will find that training is available for many other useful skills you can learn: CPR, first aid, orienteering, and search and rescue are just a few activities with active amateur involvement.

Operating in an Emergency

You hope it never happens, but what if worse comes to worst? All emergencies are different, of course, so a step-by-step procedure is not going to be very useful. Here are some solid principles to follow instead, based on the ARES Field Resources Manual.

When disaster strikes, do the following:

1. **Check that you and your family and your property are safe and secure before you respond as an emcomm volunteer.**

2. **Monitor your primary emergency frequencies.**

3. **Follow the instructions you receive from the net control or other emergency official on the frequency.**

   Check in if and when check-ins are requested.

4. **Contact your local emergency communications leader or designee for further instructions.**

Everyone is likely to be fairly excited and tense. Keep your head on straight and follow your training so that you can help rather than hinder in an emergency situation.

Reporting an accident or other incident

Reporting an accident is more common than you may think. Anybody who spends time driving can attest to the frequency of accidents. I personally use ham radio to report accidents, stalled cars, and fires. Don't assume that people with cellphones are doing it. Know how to report an incident quickly and clearly.

When you have an emergency situation to report, follow these steps if your radio has an autopatch:

1. **Turn up your radio's power to the limit and clearly say, *"Break"* or *"Break Emergency"* at the first opportunity.**

   If one station is weak, a stronger signal can get the attention of listening stations. Don't shy away from interrupting an ongoing conversation.

2. **After you have control of the repeater or the frequency is clear, state that you have an emergency to report.**

3. **State clearly that you are making an emergency autopatch and then activate the autopatch system.**

   If you can not activate the repeater's autopatch, you may ask another repeater user to activate it for you. Or, on HF or VHF, you can ask for someone to make an emergency relay to 911. In this case, report all the necessary material and then stand by on frequency until the relaying station reports to you that the information is relayed and the call is complete.

   See Chapter 9 to find out how to activate your radio's autopatch feature.

4. **Dial 911 and when the operator responds, state your name and that you are reporting an emergency via amateur radio.**

5. **Follow the directions of the operator from there.**

   If the operator asks you to stay on the line, do so and ask the other repeater users to please stand by.

6. **When the operator finishes, release the autopatch and announce that you released the autopatch.**

Whether you use a repeater's autopatch feature or relay the report by another repeater user, you need to be able to generate clear, concise information. To report an automobile accident, for example, you should know:

- The highway number or street
- The address or approximate mile marker of the highway
- The direction or lanes the accident occurred
- Whether the accident is blocking traffic
- If injuries are apparent
- If the vehicles are on fire, are smoking, or have spilled fuel

Similarly for fires and other hazards, the dispatcher wants to know where it is and how serious it appears. Don't guess if you don't know for sure! Report what you know, but don't embellish the facts.

Making and responding to distress calls

Before an emergency occurs, be sure you know how to make a distress call on a frequency where hams are likely to be listening, such as a marine service net or a wide-coverage repeater frequency. Store at least one of these frequencies in your radio's memories, if possible. Anyone, licensed or not, can use your radio equipment in an emergency to call for help on any frequency. You won't have time to be looking at net directories in an emergency. Do the following things when you make a distress call.

- ✓ **If you need immediate emergency assistance, the appropriate voice signal is MAYDAY and the appropriate Morse code signal is SOS (yes, just like in the movies).**

  Maydays sound something like: *"MAYDAY, MAYDAY, MAYDAY, this is NØAX,"* followed by:

  - Your location (latitude/longitude) or address of the emergency

  - The nature of the emergency

  - What type of assistance you need — such as medical or transportation aid

- ✓ **Repeat your distress signal and your call sign for several minutes or until you get an answer.** Even if you don't hear an answer, others may hear you.

- ✓ **Try different frequencies if you do not get an answer.** If you do decide to change frequencies, announce to what frequency you are moving so that anyone hearing you can follow.

If you hear a distress signal on the air:

- ✓ **Immediately find something to record information.** Note the time and frequency of the call. To help the authorities render assistance as quickly as possible, note the following information:

  - The location (latitude/longitude) or address of the emergency

  - The nature of the problem

  - What type of assistance he or she needs — such as medical or transportation aid

  - Any other information that is helpful

- ✓ **Respond to the call.** Say *"[Give the station's call sign], this is [your call sign]. I hear your distress call. What is your situation?"*

Using Morse code, you send *"[station's call sign] DE [your call] RRR WAT UR INFO?"* or something similar. Let the station in distress know who you are and that you hear them.

✔ **After you acquire the information, ask the station in distress to remain on frequency.**

✔ **Call the appropriate public agency or public emergency number, such as 911.** Explain that you are an amateur radio operator and that you received a distress call.

The dispatcher either begins a process of asking you for information or transfers you to a more appropriate agency.

Follow the dispatcher's instructions to the letter. The dispatcher may ask you to act as a relay to the station in distress.

✔ **As soon as possible, report back to the station in distress.** Tell them who you contacted and any information you have been asked to relay.

✔ **Stay on frequency as long as the station in distress or the authorities need your assistance.**

Supporting emergency communications outside your area

What do you do in case of a disaster or emergency situation outside of your immediate vicinity? How can you be of assistance? The best thing you can do is make yourself available to the on-site communications workers, but only if called upon to do so. Because most of the important information from a disaster flows out, not in, you don't want to get in the way.

For example, if a hurricane is bearing down on Miami, getting on the air and calling, *"CQ Miami!"* is foolish. The chance you have of actually rendering assistance are minimal and you stand a chance of diverting some actual emergency need from the proper authorities. Instead, support the communications networks that the Miami hams depend on. Check in to your NTS local nets to see if any messages need relayed to your location. Monitor the Hurricane Watch Net on 14.325 MHz and any Florida emergency net frequencies. Tune to the bands that support propagation to Florida, in case someone is calling for help.

Here's another example — say a search-and-rescue (SAR) operation in the nearby foothills is taking place with nets on 2-meter repeaters and several simplex frequencies coordinating the activities. Do you check into the SAR nets? No! But you can *monitor* (listening without transmitting) their operation to see if an opportunity arises for you to provide assistance, especially if you have beam antennas that you can aim directly at the area. (See Chapter 13 for more on antennas.) If you can set your radios to listen to the repeater input frequencies as well as the outputs, you may hear a weak station unable to activate the repeater. If you monitor the simplex frequencies, you may act as a relay station. Two stations in hilly areas may be unable to communicate directly but you can hear both and can relay communications. If such a situation occurs, you can break in and say, *"This is [your call sign] and I can copy both stations. Do you want me to relay?"*

You need to help information flow out from the disaster site, not force more in. Listen, listen, listen. That's good advice most of the time.

Providing Public Service

In between emergencies, hams perform other valuable public service in many ways. After you become associated with a local emergency communications group, you can use your ham radio skills in many opportunities for the public's benefit.

Weather monitoring

One of the most widespread public service functions is that of the amateur weather watcher. In many areas, particularly with frequent severe weather conditions, nets devoted to reporting local weather conditions meet regularly. Some nets meet once or twice every day and others only when there is a threat of severe weather.

Many weather nets are associated with the NOAA SKYWARN program (www. skywarn.org). Groups reporting weather conditions under the SKYWARN program feed information to the National Weather Service (NWS), which uses the reports in forecasting and severe weather management. In some areas, a net control station may operate a station from the NWS itself. For information on whether a SKYWARN net is active in your area, follow the Local Skywarn Groups link on the SKYWARN home page or enter **SKYWARN net** into an Internet search engine.

Other weather nets may operate on VHF/UHF repeaters or on 75-meter voice nets. For example, the New England Weather Net meets on 3905 kHz every day at 1030Z. Informal weather nets on local repeaters are common. Ask around to see if one operates in your area. They are usually active at commuting drive times.

Parades and sporting events

Amateurs participate in parades and sporting events to provide the event managers with timely information and coordination. In return, amateurs get good training in communications procedures and operations that simulate real-life emergencies. For example, you can think of a parade as similar to a slow-speed evacuation. A lost-child booth at a parade is similar to a small search-and-rescue operation. Helping keep track of race entrants in a marathon is good practice for handling health-and-welfare messages.

A lead representative of the amateur group usually coordinates plans with the event management. The group then deploys whatever the plan requires. Depending on the size of the event, all communications may take place on one simplex frequency or several may be required. Information may be restricted to simple status or actual logistics information may be relayed. Communications support includes a wide variety of needs.

Event managers typically work with a single club or emergency services group that manages the ham radio side of things. If you want to participate in these events, start by contacting your Section Manager. The SM directs you to one or more individuals active in public service who can let you know about upcoming events.

When you support an event, be sure to get the appropriate identification. Dress similarly to the rest of the group and obtain any required insignia. Have a copy of your amateur license and some photo ID. Make sure you take water and some food with you in case you are stationed somewhere without support. Don't assume you'll be out of the weather and protect yourself against the elements. Have your identification permanently engraved or attached to your radio equipment to protect against theft.

Operating on Nets

Nets are one of the oldest ham radio activities. The first net was probably formed as soon as two hams were on the air! Nets are just regularly scheduled on-the-air meetings of hams with common interests. Sometimes the interests are strictly for pleasure, such as collecting things, playing radio chess, or pursuing the Worked All States awards. Other nets are more utilitarian, such as those for traffic handling, emergency services, weather reporting, and so forth.

Nearly all nets have a similar basic structure. A Net Control Station (NCS) initiates the net operations, maintains order, directs the net activities, and then terminates net operations in an orderly way. Stations wishing to participate in the net check in at the direction of the NCS. A net manager defines net policy and focus, and works with the NCS stations to keep the net meeting on a regular basis.

Information, such as a formal radiogram or just a verbal message, is exchanged during a net either on frequency or off frequency. Nets primarily for discussion of a common interest or for selling and trading equipment tend to keep all the transmissions on one frequency so that everyone can hear them. This system is quite inefficient for a traffic handling or emergency net, so the NCS sends stations off frequency to exchange the information and then return to the net frequency.

Here's an example of an NCS directing an off-frequency message exchange during an NTS traffic-handling net:

> W2——: *I have one piece of traffic (a single message) for Baltimore.* (W2—— either received the message on another net or can be the originating station.)
>
> NCS: *W2—— standby. K3—— can you take that traffic for Baltimore?* (K3—— registered earlier *(checked in)* with the NCS and reported his location as somewhere near Baltimore.)
>
> K3——: *I can take the Baltimore traffic.* (K3—— accepts and delivers the message to the addressee.)
>
> NCS: *W2—— and K3—— move 5 down and pass the traffic.*

This transmission means W2—— and K3—— are to tune 5 kHz below the net frequency, re-establish contact, and W2—— then transmits the message to K3——. When they are done, both return to the net frequency and report to the NCS. K3—— may stay on frequency until the net finishes or may immediately leave to deliver the message.

If you want to *check in* to a net, you register your call sign and location or status with the NCS. Be sure you can hear the NCS clearly and that you can understand his or her instructions. If you're not a regular net member, wait until the NCS calls for visitors. When you check in, give your call sign once, phonetically if on phone. If the NCS doesn't copy your call sign the first time, repeat your call sign or the NCS requests a relay from one of the listening stations that may hear you.

You can check in with business (such as an announcement) or traffic (messages) for the net in a couple of ways — listen to the net to find out which is the appropriate method. The most common method is to state when checking in, *"NØAX with one item for the net."* The NCS acknowledges your item and you then wait for further instructions. Or you can just check in with your call sign and when the NCS acknowledges you and asks if you have any business for the net, reply, *"One item."* Listen to other net members checking in and when in Rome, check in as the Romans do.

If you want to contact one of the other stations checking in, you can either declare this intention when checking in as if it were net business or you can wait until the check-in process is complete and the NCS calls for net business. Either way, the NCS asks the other station to acknowledge you and puts the two of you together following net procedures.

Nets are run using many different methods. Some methods are very formal and others are more like an extended round-table QSO. The key is to listen, identify the NCS, and follow the directions. The behavior of other net members is your guide.

Handling Traffic

The National Traffic System (NTS), managed by the ARRL Field Organization, is the backbone of all amateur message handling in North America. The NTS runs every day, during emergencies and normal times alike. Traffic handling, while less popular an activity than in years past, builds skilled and accurate operating techniques, making you ready for any emergency.

Is traffic handling still needed?

Does traffic handling sound sort of antique? It should, being one of the oldest activities in ham radio. Why, in this age of wireless LANs and coast-to-coast links, does traffic handling still exist? The main reason is that when all else fails — and sophisticated communications systems such as cell phones and the Internet fail quickly in a disaster or emergency — ham radio traffic handlers fill the gap in an accurate and accountable way until those faster systems are brought back online. Under such circumstances, large numbers of simple text messages must be sent quickly and reliably. For example, health and welfare messages stream in and out of the afflicted areas until more sophisticated systems are brought in or restored. Even when digital means of transmission are available, the same type of message structure is often employed to make the process accountable and traceable as good emergency management requires. *Remember:* Traffic handling is supposed to work in an emergency, when the chips are down and only the barest minimum of resources may exist.

Traffic consists primarily of text messages in the form of *radiograms* that are very similar to old-style telegrams. Each message is relayed from ham to ham using time-tested procedures until the message reaches its destination where it's delivered by an appropriate method. For detailed information on traffic handling, the *ARRL Operating Manual* has an extensive chapter on handling traffic procedures and jargon.

Figure 10-2 shows how messages entering the NTS start with nets at the local level and are then passed up to section, regional, and area nets. Wide-area nets connect the various regions as the messages move closer to their destinations. Moving back down the tree, the messages are delivered to section nets and finally, local nets. These NTS nets are managed by a Section Traffic Manager (STM) who is part of the ARRL Field Organization.

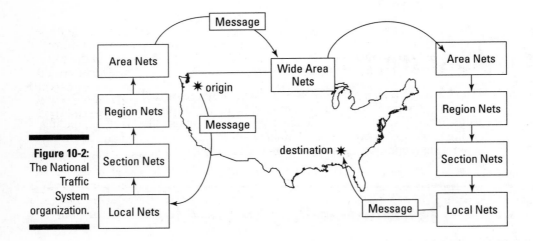

Figure 10-2: The National Traffic System organization.

Handling traffic is an excellent way for young hams to gain experience and take on responsibility. You don't have to be a certain age, only have a desire to perform well and pay attention to detail and procedure. You can participate no matter where you live and whether you have just an hour or two a week or can check in twice a day. The more volunteers available, the better the chances of being able to support emergency operations.

Visual or physical handicaps are no obstacle either. When I was active in Missouri nets, blind ham Ruth KØONK was the top traffic handler in the state month after month with hundreds of messages to her credit.

If you develop a taste for traffic handling, you can request that your Section Manager appoint you as an Official Relay Station (ORS). The ORS appointment signifies that you're developing your skills in traffic handling and can use them to support your community in an emergency. You may choose to specialize in traffic handling on one mode, become a net manager, or act as a liaison between regions.

Getting started

If you decide to give traffic handling a try, your best bet is to find an NTS local net in your area. Go to the ARRL Net Search Web page (www.arrl.org/ FandES/field/nets/client/netsearch.html), select the Local Nets radio button, your state from the U.S. State drop-down list, and the National Traffic Forwarding Nets Only option from the National Traffic Affiliated drop-down list. You get a list of local nets in your state that handle traffic. Click the net name to find out more about the net.

On this Web site, you can also find a number of training nets using both voice and Morse code. Follow the same search process, but select the State Nets radio button, select your state from the list, and enter **Training** in the Net Name box. If you do not see any training nets for your state, search again with the Net Name box blank and review the resulting list. If you are unsure of which nets are suitable, click the link for the state emergency net and contact the net manager listed there for advice. You can also contact your Section Manager — follow the link for your section at www.arrl.org/sections/?sect=.

When you select a net, listen in for a session or two. Determine when they ask for visitors and how member stations check in. After you understand the net procedures, check in as a visitor and ask for an after-net contact with the NCS. During that contact, you can ask for information about the net and whether any net guidelines are available. Then start checking in on a regular basis.

Handling a piece of traffic

The magic day arrives and the NCS hooks you up with another station with a radiogram for your town. Don't be afraid to say, *"This is my first message, please take it slow."* An experienced traffic handler takes all the time you need to get the job done right. To prepare for this day, download and review the radiogram guidelines FSD-218 at www.arrl.org/FandES/field/forms (you need to scroll down the page before you come across it).

If you have a pad of radiogram forms (shown in Figure 10-3), so much the better, but they're not required. The radiogram guidelines describe each bit of information just as it comes in from the traffic system. All you have to do is write down the information and follow the instructions.

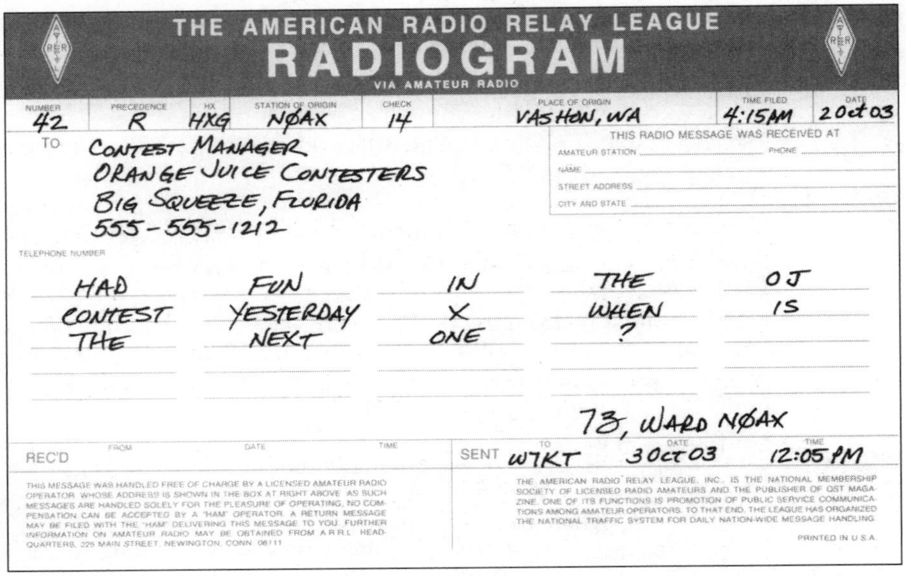

THE AMERICAN RADIO RELAY LEAGUE

RADIOGRAM

VIA AMATEUR RADIO

NUMBER	PRECEDENCE	HX	STATION OF ORIGIN	CHECK	PLACE OF ORIGIN	TIME FILED	DATE
42	R	HXG	NØAX	14	VASHON, WA	4:15AM	20ct 03

TO CONTEST MANAGER
ORANGE JUICE CONTESTERS
BIG SQUEEZE, FLORIDA
555-555-1212

TELEPHONE NUMBER

THIS RADIO MESSAGE WAS RECEIVED AT
AMATEUR STATION ______ PHONE ______
NAME ______
STREET ADDRESS ______
CITY AND STATE ______

HAD	FUN	IN	THE	OJ
CONTEST	YESTERDAY	X	WHEN	IS
THE	NEXT	ONE	?	

73, WARD NØAX

REC'D	FROM	DATE	TIME	SENT	TO	DATE	TIME
					W7KT	30Oct03	12:05PM

THIS MESSAGE WAS HANDLED FREE OF CHARGE BY A LICENSED AMATEUR RADIO OPERATOR WHOSE ADDRESS IS SHOWN IN THE BOX AT RIGHT ABOVE. AS SUCH MESSAGES ARE HANDLED SOLELY FOR THE PLEASURE OF OPERATING, NO COMPENSATION CAN BE ACCEPTED BY A "HAM" OPERATOR. A RETURN MESSAGE MAY BE FILED WITH THE "HAM" DELIVERING THIS MESSAGE TO YOU. FURTHER INFORMATION ON AMATEUR RADIO MAY BE OBTAINED FROM A.R.R.L. HEADQUARTERS, 225 MAIN STREET, NEWINGTON, CONN. 06111

THE AMERICAN RADIO RELAY LEAGUE, INC., IS THE NATIONAL MEMBERSHIP SOCIETY OF LICENSED RADIO AMATEURS AND THE PUBLISHER OF QST MAGAZINE. ONE OF ITS FUNCTIONS IS PROMOTION OF PUBLIC SERVICE COMMUNICATIONS AMONG AMATEUR OPERATORS. TO THAT END, THE LEAGUE HAS ORGANIZED THE NATIONAL TRAFFIC SYSTEM FOR DAILY NATION-WIDE MESSAGE HANDLING.

PRINTED IN U.S.A.

Figure 10-3: An ARRL radiogram form with a sample message.

A radiogram consists of these parts:

- **Preamble:** The first batch of information you receive, the preamble is information about the message — not the message itself. It describes the nature and origin of the message so that you can handle it and reply, in case you need to do so. The preamble improves the message accuracy.
- **Addressee:** Along with the name, the addressee information usually has a street address and a telephone number. Some messages only have a name, town, and telephone number. You deliver the message by telephone or whatever means are available.
- **Text message:** Radiograms are intended to be 25 words or less, so you don't have to deal with a lot of text. Be sure you have the spelling correct on any unusual words. The only punctuation you're likely to encounter is an X in place of a period and a ? at the end of a question. Don't be confused by the *check* of the message, which is just a count of word groups, number groups, and mixed groups (letter-number combinations, such as a serial number) in the text portion only. The check is to allow you to count through the text and verify that you received the right amount of information. Don't be afraid to ask the transmitting station if you're not sure about counting the text's elements.

  The text may include something like ARL FORTY SIX. This text refers to ARRL Numbered Radiograms, which are special condensations of common groups of words. ARL FORTY SIX (they're always spelled out) means "Greetings on your birthday and best wishes for many more to come." These are quite a time saver! A complete list (FSD-3) is available from the ARRL Field Organization Forms page (www.arrl.org/FandES/field/forms).

✔ **Signature:** The name or identification of the message's originator.

Take your time and carefully check the entire message to be sure you copied it correctly. Check with the transmitting station if you have any questions. When you're sure, you can say, *"QSL!"* (meaning "received completely" with pride and report back to the NCS.

Delivering the message

Delivering a message is a pretty easy part of traffic handling. All you have to do is call the addressee and say, *"Hello, my name is — , I'm an amateur radio operator and I have a message for you from — ."* You need to let them know you're not a telemarketer and that you have a message from someone they are likely to know.

After your addressee is ready to receive the message, carefully read it back to him or her (just the text, not the preamble or addressee). Convert any numbered radiogram to its equivalent text — even if you're delivering the message to another ham. When you're done, be sure everything is okay and ask if he or she has a reply.

If so, be sure to get the complete addressee information (be sure of the spelling) and try to limit the response to 25 words or less. You can use the numbered radiograms to save space, too. Note the local time and date when you accept the message. Assign it a precedence, add any handling instructions (which are optional), and count up the words.

After you write the message, take it back to the next net session and when you check in, say *"[your call] with one message!"*

Sending a message

When you get back to the net and the NCS assigns you to a station that accepts your traffic, ask for help in determining the message's *check,* or word count (if you need help). Tell the station the check you came up with and ask him or her to help you confirm it. Before the message goes on its way, the check needs to be correct or subsequent relaying stations will think an error is in the message and it may come back to you.

With the complete message ready to go, slowly read each portion of information to the relaying station using the following steps:

1. **Start with the preamble and then say, *"Break for addressee."***

   The relaying station either says *"Go ahead"* or asks for clarification about some item in the preamble.

2. **Read the addressee information and say, *"Break for text."***

3. **Read the text and signature and say, *"End of Message."***

   If the relaying station didn't get the message all correct, he or she lets you know and asks for a *fill* or repeat. For example, they may say, *"All after feldspar,"* which means "Repeat everything after the word feldspar."

Soon, the relaying station reports *"QSL!"* and you've done it!

Chapter 11

Specialties

· ·

In This Chapter

▶ Contacting distant stations

▶ Participating in contests

▶ Pursuing awards

▶ Operating with low power

▶ Exploring the digital modes

▶ Using satellites

▶ Transmitting images

· ·

*A*fter you get rolling with casual operating, you can begin to explore a whole world of interesting specialties within ham radio. Specialties, to many, are the real attraction of the hobby.

In this chapter, I give you an overview of the most popular activities, cover some of the basic techniques and resources, and demystify a little bit of the specialized jargon ham radio seems to attract. I don't hit all of the possible activities by any means, but start with the activities I discuss in this chapter and you'll discover many others along the way, especially if you read the magazines and browse the popular Web sites.

DX-ing

Right after traffic handling (Chapter 10), pushing your station to make contacts over greater and greater distances (DX means *distant stations*) is the second oldest activity in all of ham radio. Somewhere out in the ether, a station is always just tantalizingly out of reach and the challenge of contacting that station is the purpose of DX-ing.

Thousands of hams across the continents and around the world like nothing better than to make contacts (QSOs) with someone far away. These hams seem to ignore all nearby stations. Their logs are filled with exotic locations. Ask them about some odd bit of geography and you are likely to find that they not only know where it is, but some of its political history and the call sign of at least one ham operator there. These hams are DXers.

The history of ham radio is tightly coupled to DX-ing. As transmitters became more powerful and receivers more sensitive, the distances a station could make contact were a direct measure of quality. Hams quickly explored the different bands and follow the fluctuations of the ionosphere. DX-ing drives improvements in many types of equipment.

Today, intercontinental contacts on the HF frequencies traditionally considered to be the shortwave bands are common but still thrilling. Cross-continental contacts on VHF and UHF once thought impossible are made in increasing numbers. Because the sun and the seasons are always changing, each day you spend DX-ing is a little (and sometimes a lot) different. Sure, you can log on to an Internet chat room or send e-mail around the world, but, like fishing, logging a QSO in the log, mastering the vagaries of the ionosphere, and getting through to a distant station is a real accomplishment non-hams can never know.

DX-ing on the shortwave bands

The following sections show you how to use the shortwave or HF bands to "work DX." These are the bands with frequencies below 30 MHz on which signals routinely travel all over the Earth, bouncing between the ionosphere and the Earth's surface as they go. The ease with which signals can be exchanged between continents on the shortwave bands attracts many thousands of adherents. Because the signals propagate so widely, HF DX-ing is often a worldwide event with stations calling from several continents. VHF and UHF DX-ing is no less exciting, but requires a different approach, which is covered later. Propagation is much more selective, so fewer signals are heard at one time and are usually from stations concentrated in a few areas. These differences make what is fundamentally the same pursuit — making contact as far away as you can — quite different on the shortwave bands versus VHF and UHF.

For more information on propagation I recommend checking out the technical information available on the *Ham Radio For Dummies* Web site to find out how the ionosphere affects long-haul contacts for HF, VHF, and UHF signals. In-depth reading about shortwave DX-ing techniques can be found in *The Complete DXer* by Bob Locher W9KNI (published by Idiom Press). Now in its third edition, Bob's book has Elmered legions of beginning DXers.

Tuning for DX

Before starting out, you need to know that even if you have a very modest home or mobile HF station, you can work DX. Skill and knowledge compensate for a great deal of disparity in equipment. Nowhere is this concept more true than in hunting the elusive DX. The first skill to learn is not how to transmit, but how to listen.

When working DX, in all cases, start at the bottom of the band or as close as your license privileges permit. As I discuss in Chapter 9, the best DX tends to collect there. Stop at each signal along the way, even those that sound like casual contacts, to determine who is on the air. Listen for obvious accents and signals with a curious, hollow, or fluttery sound.

Signals coming from far away have to make several hops off the ionosphere — sometimes as many as five or six! — to get to your antenna. These hops divide the signal into multiple paths that have slightly different travel times. The paths interfere with each other as they arrive at your antenna, smearing the signal out in time and making its strength change rapidly. Learn to recognize that sound because, for sure, it means DX is at hand!

Check the frequencies frequented by DX-peditions — trips lasting a few days to exotic and desirable locations made specifically to *activate* (make contacts from) them for DXers around the world. You can usually find these adventurers on frequencies in the lower 25 kHz of the CW bands (the Extra Class segments) and at the high end of the Extra Class segments on phone. Common frequencies include 3.795, 14.195, 21.295, and 28.495 MHz.

Program the popular DX frequencies into your rig's memory for easy access.

Keep tuning and listening, noting what you hear and at what times. When DX-ing, experience with the characteristics of a band's propagation is the best teacher. Try to detect a pattern when signals from the different population centers appear and how the seasons affect propagation on the different bands. Soon you recognize the signals of regulars on the band, too. As usual, the key is to listen, listen, listen.

If you plan on doing a lot of DX-ing, purchase a copy of the ARRL DXCC List (www.arrl.org/awards/dxcc) and a ham radio prefix map of the world for reference. These tools help you figure out what country corresponds to the call signs you hear. You sometimes hear special or unusual call signs. With an ITU prefix list, you can figure out those call signs' countries of origin. A very detailed prefix-country list is available online at www.ac6v.com/prefixes.htm#PRI.

While you're collecting resources, here's another suggestion: Centered on your location, an *azimuthal-equidistant* or *az-eq* map, such as the one in Figure 11-1, tells you what direction a signal's coming from. Because signals travel along the "Great Circle" paths between stations (imagine a string stretched tightly over a globe between the stations), the path for any signal that you hear is along the radial line from the middle of the map (your location) directly to the other station. If the path goes the "long way around," it goes off the edge of the map (which is halfway around the world from your station) and reappears on the other side. This is the "long path." Most signal paths stay entirely on the map because they take the "short path." Some az-eq maps are available in the *ARRL Operating Manual* and you can also generate one online from several sources, including www.wm7d.net/azproj.shtml.

Daytime DX-ing

You must account for the fluctuations in the ionosphere when you're DX-ing. Depending on the hour, the ionosphere either absorbs a signal or reflects it over the horizon. In the daytime, the 20, 17, 15, 12, and 10-meter bands, called the *High Bands,* tend to be "open" (support propagation) to DX stations. Before daylight, signals begin to appear from the east, beginning with 20-meters and progressing to the higher bands over a few hours. After sunset, the signals linger from the south and west for several hours with the highest frequency bands closing first in reverse order. Daytime DXers tend to follow the *Maximum Useable Frequency (MUF),* the highest signal the ionosphere reflects. These reflections are at a very low angle and so can travel the longest distance for a single reflection (one reflection is called a *hop*) and have the highest signal strengths.

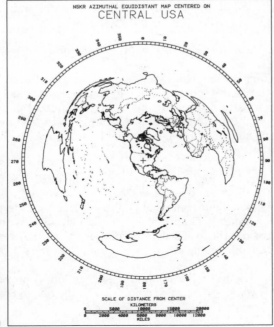

Figure 11-1: An azimuthal-equidistant map centered on the Midwest United States.

Nighttime DX-ing

From 30-meters down in frequency are the nighttime bands of 30, 40, 60, 80, and 160-meters, known as the *Low Bands*. These bands are throttled during the daytime hours by absorption in the lower layers of the ionosphere. After the sun begins to set, these bands start to come alive. First, 30, 40, and 60-meters may open in late afternoon and stay open somewhat after sunrise. 80 and 160-meters, however, make fairly rapid transitions around dawn and dusk. Signals between stations operating on 80 and 160-meters often exhibit a short (15 to 30 minute) peak in signal strength when the easternmost stations are close to sunrise. This is known as the *dawn enhancement*. This time is good for stations with modest equipment to be on the air and take advantage of the stronger signals on these more difficult DX bands.

160-meters is known as *Top Band* because it has the longest wavelength of any current amateur band. This long wavelength requires larger antennas. Add in more atmospheric noise than at higher frequencies and you have a challenging situation. That's why some of the most experienced DXers love Top Band DX-ing. Imagine trying to receive a 1 kilowatt broadcast station halfway around the world. That's what the Top Band DXer is after! As difficult as this task sounds, many of the top DXers have managed it.

Contacting a DX station

Making a call to a DX station requires a little more attention to the clarity of your speech and sending than making a call to a nearby ham. *Remember:* Your signal likely has the same qualities as the DX station — hollow or fluttery and weak — so speak and send extra carefully. Give the DX station's call sign using the same phonetics they are using and then repeat yours at least twice, using standard phonetics. On Morse code contacts, send the DX station's call sign once and your call sign two or three times at a speed matching that of the DX station.

DX contacts, except when signals are quite strong, tend to be shorter than contacts with nearby stations. When signals are very weak or a station very rare, a contact may consist of nothing more than a confirmation that you each have the call signs correct and a signal report as described in Chapter 8. To confirm the contact, both you and the DX station must get each other's call signs correct. To do that, use standard phonetics (on voice transmissions), speak clearly, and enunciate each word. New hams often don't realize that multiple hops and skips around the world have a pretty dramatic effect on speech intelligibility, none of it for the better. Speak relatively slowly, don't slur your words or mumble, and keep your transmissions short.

When it's time to conclude the contact, you need to let the other station know if you will be sending a *QSL card* (these cards are commonly known as *QSLs*) to confirm that the contact occurred. Collecting these cards, like those in Figure 11-2, is a wonderful part of the hobby. See Chapter 14 for more info on QSL cards.

Tracking the sun

Because the sun is so important in determining what bands are open and in what direction, you need to know what portions of the Earth are in daylight and darkness. You can use a number of tools to keep track of the sun. The very handy DX Edge, once made by Xantek and currently only available at hamfests and on the various used-equipment-for-sale Web sites, is a pocket-sized world map with sliding overlays that show the light and dark areas of the Earth in different seasons. Many DXers keep one at their elbow when chasing DX. You can substitute the online gray line map instead (found at `dx.qsl.net/ propagation/greyline.html` and shown in the figure below). A useful standalone software package that has many types and styles of real-time world and regional maps is Geoclock (`home.att.net/~geoclock`).

Figure 11-2: DX QSL cards from New Zealand, Japan, Pratas Island, and England.

YOU DON'T NEED TO SHOUT INTO THE MICROPHONE! Shouting doesn't make you any louder at the other end! By adjusting your microphone gain and speech processor, you can create a very understandable signal at normal voice levels! Your contacts and family will thank you for doing so. Save the shouting for celebrating your latest DX contact!

If you call and call and can't get through or if the stations you contact ask for a lot of repeats and fills (in other words, if they often ask you to repeat yourself), you probably have poor audio quality. Have a nearby friend, such as a club member, meet you on the air when the bands are quiet and do some audio testing. Check to see if you have hum or noise on your audio. Noise is often the result of a broken microphone cable connection, either in the microphone itself or at the radio connector. You may not be able to tell you have a problem from the radio's power meter output, so an on-the-air check is necessary to find it. Inexpensive, old, and non-communications microphones (such as computer microphones) often have poor fidelity. If your on-the-air friend says you sound like a bus station PA system, upgrade to a better microphone!

Pileups

Aptly named, a pileup is just that: a pile of many signals trying to get through to a single — often quite rare — station. Pileups can sound like a real mess, but if you listen carefully, you'll notice that some stations get right through. How do they do this? They listen to the rare station's operating procedure, find what kind of signals the operator is listening for, and carefully time their call. If they don't get through the first time, they stop calling and listen until they have it figured out. These smooth operators use their ears instead of their lungs or amplifiers to get through. You can, too, by listening first and transmitting second. Here are some common tricks to listen for and try yourself:

- ✔ Time your call a little bit differently than everybody else. Wait for a second or two before beginning or wait for the short lull when most have given their call once and are listening.

- ✔ Make your signal sound a little bit differently, higher or lower by offsetting the transmit frequency 200 or 300 hertz.

- ✔ Just give your call once or twice before listening for the DX station. Some folks never seem to stop calling — how would they know if the DX station did answer them?

- ✔ Use phonetics similar to others that have gotten through.

Try to figure out what the DX station hears well and try to do that.

Splits

The term *split* is used when a station is transmitting on one frequency, but listening on another. This procedure is common when many stations are trying to get through to a single station, such as a rare DX station. You can tell that a station is *working split* when you hear the station contacting other stations but you can't hear those stations' responses. It can also work the other way around: Sometimes you tune in a pileup of stations trying to contact a DX station — but you aren't able to hear the DX station's responses. Typically, the DX station's split listening frequency is a few kHz above the transmitting frequency. The station being called gives instructions such as *"listening up two"* or *"QRZed 14205 to 14210."* The former means he's listening for stations 2 kHz above his transmit frequency. The latter means he's listening in the range between the two frequencies, in this case probably 14.205 to 14.210 kHz.

To "work split," set one of your radio's main tuning controls (called *VFOs,* which is an abbreviation of *variable frequency oscillator*) to the transmit frequency and the other to the receive frequency. These are usually referred to as "VFO A" and "VFO B." Set up the radio to listen where one VFO is set and transmit on the other. Read your rig's manual carefully so that you can do this properly. If you accidentally transmit on the wrong VFO, you interfere with the DX station. This method is the easiest way of working split, but it's not the only one: You can also establish the transmit and receive frequencies as memory selections on your radio, and then jump between memories. Don't bother trying to dial back and forth — you won't be quick enough. With either method, practice for a while first, before trying it out on the air.

Using (and abusing) DX spotting systems

DXers share the frequencies and call signs of DX stations they discover on the air through an extensive, worldwide system of VHF packet radio networks and Web pages. This system is called *spotting,* and the message that describes where you can find the DX station is called a *spot.* The following is an example from the popular Web site www.dxsummit.com:

> W5VX 7003.7 A61AJ 0142 05 Nov

This spot means that W5VX is hearing A61AJ from the United Arab Emirates (A6 is the prefix of call signs for amateurs in the UAE) on a frequency of 7003.7 kHz at 0142Z (01:42 AM in London) on November 5th.

If you live in any populated area of North America or Europe, you probably can access one of the VHF packet radio DX spotting systems. If you have Internet access, you can also log on to the DX spotting system Web sites using a browser or by using TELNET (a text-only Internet communications program included with most computer operating systems) to log on to local "packet clusters," which are linked to the VHF packet systems. (See the section "Packet," later in this chapter, for more on packet systems.) Numerous DX Web sites and clusters are listed at http://www.ac6v.com/dxlinks.htm.

Although jumping from spot to spot can be a lot of fun, maintaining your tuning and listening skills is important. Be sure that you have the station's call sign correct before you put it in your log — contacting what you think is a rare station, only to find out that due to a *busted spot,* your fabulous DX contact wasn't so fabulous after all, is a disappointment. Because spotted stations attract quite a crowd, working DX sometimes by not chasing the spotted stations and tuning for them yourself may be easier. Don't become dependent on the spotting networks.

Getting awards

Many DX-ing award programs, and the most popular are listed in Table 11-1, are suited to widely varying levels of interest. If you are just getting started, the Worked All Continents (WAC) award is for you. What you need is a contact with each of the six populated continents: Europe, Africa, Asia, North and South America, and Oceania (islands in the South Pacific and Australia). No matter where you live around the world, one continent remains aloof, tantalizing you until you break through and make the contact. The Worked All States (WAS) program is popular both in and out of the Unites States and is a great way to get used to hunting elusive contacts. My last state was Vermont. What will yours be?

Table 11-1	Popular DX Award Programs	
Sponsor	*Award Program*	*Achievement*
ARRL (www.arrl.org/awards)	Worked All Continents (WAC)	Make a contact in each of six continents.
	Worked All States (WAS)	Make a contact in each of the 50 U.S. states.
	DX Century Club (DXCC)	Make a contact with 100 of the DXCC entities (currently 337 countries, islands, and territories are on the list).
CQ Magazine (www.cq-amateur-radio.com/awards.html)	Worked All Zones (WAZ)	Contact all 40 of the world's CQ-defined zones, which are different regions of the world, such as Eastern Europe, Japan, and so on.
	Worked Prefixes (WPX)	Several types of awards for contacts with different types of call signs.
Radio Society of Great Britain (RSGB) (www.rsgb.org)	Islands on the Air (IOTA)	Various levels of awards for contacting saltwater islands around the world.

The other awards in Table 11-1 are more challenging. For example, DXCC — the undisputed leader of DX award programs — is awarded when the applicant has confirmed contacts with 100 entities (countries, colonies, islands) around the world. Because the awards are challenging, their achievement does signify a worthwhile accomplishment. As a result, making contacts in pursuit of the awards is a very popular segment of ham radio.

Most DX awards programs reward achievement in the same manner: First, you must qualify for the basic award (100 entities, 100 islands, 300 prefixes, and so on). You receive a certificate and your first *endorsement* (a sticker or other adornment that signifies a specific level of achievement). From that point, you can receive additional endorsements for higher levels of achievement: more contacts on one band, more contacts in one geographic region, and so on. Because these levels of success are so open-ended, you can achieve them all. (For more on awards, see the section "Chasing Awards," later in this chapter.)

DX-ing on the VHF/UHF bands

Although DX-ing on the traditional shortwave bands is very popular, an active and growing community enjoys DX-ing on the bands above 30 MHz. The excitement of extending your station's capability to these bands is being shared by more hams than ever before. The explosion in popularity of VHF/UHF DX-ing is similar to the explosion of HF DX enthusiasm in the 1960s when top-quality equipment became available to the average ham. These days, the recent generation of all-band HF/VHF/UHF radio equipment puts top-notch DX-ing on the shack desktop.

With the exception of the 6-meters band — known as the Magic Band because of its sudden and dramatic openings where distant stations suddenly "appear" — these higher frequencies usually do not support the kind of long-distance, transoceanic contact common on HF because the ionosphere cannot reflect those signals. VHF/UHF DXers look for contacts using different methods of propagation.

VHF/UHF bands have unfairly gotten a reputation as being limited to line-of-sight contacts. This reputation is due to the limitations of previous generations of relatively-insensitive equipment and the prevalence of FM, which takes considerably more signal strength to provide a signal quality equivalent to SSB and Morse code transmissions. By taking advantage of well-known modes of radio propagation, you can extend your VHF/UHF range dramatically beyond the horizon.

Finding and working VHF/UHF DX

As on the HF bands, the DX is found at the very lowest frequencies on the band in the so-called weak signal segments. For example, on the 6-meters band, 50.0 to 50.3 MHz — a 300 kHz segment as large as most HF bands — is where the Morse code and single sideband calling frequencies are located. Similar segments exist on all VHF/UHF bands through the lower microwave frequencies.

When you're DX-ing on VHF/UHF, stay close to these calling frequencies or set your radio to scan across the lower end of the band and leave the radio on. Propagation between widely separated points is often short-lived. If you wait for somebody to call you or e-mail you with news about a DX station, you're probably going to miss the boat. Set your squelch control (squelch mutes the receiver unless a signal exceeds a preset level) so that the radio is just barely quieted. If anything shows up on frequency, the radio springs to life. This way, you (and whoever else is in earshot) do not have to listen to receiver hiss and random noise.

For this type of DX-ing, I recommend a small beam antenna. (Beam antennas are discussed in Chapter 12.) These antennas are easy to build, being relatively small compared to HF antennas, and are terrific home-brewing projects. Mount the antenna for *horizontal polarization* with the antenna elements parallel to the ground. You should be able to point the antennas in any horizontal direction, because signals may appear from nearly any direction at any time.

Sporadic-E

The term *sporadic-E* refers to an interesting property of one of the lower ionospheric layers — the E-layer. Somewhere around 65 to 70 miles above the Earth, illumination of the E-layer by the sun produces small highly-ionized regions that are highly reflective to radio waves. So reflective, in fact, that they can reflect 6-meter, 2-meter, 1.25-meter, and (rarely) 70-cm band signals back to Earth. These regions don't last more than an hour or two and drift around over the Earth's surface. While they're there, though, hams can use them as big radio reflectors. Their temporary nature has led to the name *sporadic-E* for the propagation paths that use them.

Sporadic-E, or *Es,* propagation occurs all through the year, but is most common in the early summer months and winter months. When Es are present, signals appear to rise out of the noise over a few seconds as the ionized patch moves into position between stations. The path may last for seconds or for an hour, with signals typically very strong in both directions. Working Es with only a few watts and very simple antennas is possible. Most VHF/UHF DXers get their start working Es openings on 6-meters, and certainly more people are actively DX-ing in that way than in any other.

Aurora

Along with sporadic-E clouds, you can find another large ionized structure in the ionosphere — the aurora! The aurora is oriented vertically, instead of horizontally like sporadic-E, but still reflects signals very well. When strong aurora is present, VHF and UHF signals are reflected over a very wide area.

One of the neatest things about auroral propagation is the ability to add its own audible signature to the signals it reflects! If you've ever seen the aurora, you understand how dynamic it is, twisting and shimmering from moment to moment. This movement is even more pronounced to radio waves. The result is that signals reflected by an aurora have a characteristic rasp or buzz impressed on the Morse tone or the spoken voice. Very strong aurora can turn Morse transmissions into bursts of white noise and render voices unintelligible. Like seeing the aurora, after you hear the auroral signature, you'll never forget it!

Tropospheric

Also known as *tropo,* tropospheric propagation occurs in the atmospheric layers closest to Earth's surface, in an area known as the troposphere. Any kind of large-scale abrupt change in the troposphere, such as temperature inversions or weather fronts, can act as a conduit for VHF, UHF, and even microwave signals over long distances. If your region has regular cold or warm fronts, you can take advantage of them to reflect or convey your signals.

Tropospheric propagation supports surprisingly regular communications on 2- and 1.25-meters and between stations in California and stations on the slopes of Hawaiian volcanoes. A stable temperature inversion layer forms over the eastern Pacific most afternoons, so a properly located station on the slope of a volcano at the right altitude can launch signals along the inversion. As the inversion breaks up near land, the signals disperse and are received by the mainland amateurs. When conditions are right, the mainlanders can then send signals back along the same path, more than 2,500 miles!

Meteor scatter

The most fleeting reflectors of all result from the tens of thousands of meteors that enter the Earth's atmosphere each day. Traveling at thousands of miles per hour, the friction as the meteors burn up ionizes the gas molecules for several seconds. These ionized molecules also reflect radio signals, so two lucky stations with the meteor trail between them can communicate for a few seconds. The ionized trails reflect radio waves for shorter and shorter durations as frequency increases. This makes the lowest frequency VHF band, 6-meters (50 MHz), the "easiest" band for beginners to make contacts using meteor scatter.

Stations that attempt to make contact in this way are called *ping jockeys,* because the radio waves bouncing of the trail have a characteristic ping sound. As you may imagine, ping jockeys go into high gear around the times of meteor showers, large and small. Because of meteor scatter, hams can enjoy meteor showers even during daylight hours!

In order to make contact during these few seconds, hams communicate very quickly. Voice operators make very short transmissions of their call signs and locations. Morse code operators send short bursts of high-speed code. Taking advantage of the capabilities of a computer and sound card, an ultra high-speed Morse code program, HSMS (for High-Speed Meteor Scatter), automates some of the more difficult aspects of this interesting mode.

For more information on meteor scatter, you can find several useful links, as well as a bulletin board for meteor scatter propagation, at dxworld.com/hsms.html.

Mountaintopping

What happens when all of the popular DX-ing methods fail to provide you with an over-the-horizon path? Well, then, move your horizon! Because VHF/UHF radios are light and antennas small, you can drive, pack, or carry your gear to the tops of buildings, hills, ridges, fire lookouts, and yes, even mountaintops.

The higher the elevation of your station, the farther your signal travels without any assistance from the ionosphere, weather, or interplanetary travelers. Camping, hiking, and driving expeditions can take on a ham radio aspect, even if you're just taking a hand-held radio. From the tops of many hills, you can see for many tens of miles and a radio can see even better than you! These expeditions are particularly popular in VHF contests, discussed in the section "Taking Part in Radio Contests," later in this chapter. All you have to do is pick up a book of topographic maps of your state, load up the car with your radio gear, and head out!

VHF propagation resources

To use the types of propagation modes I describe in this chapter effectively, you can benefit greatly from the experiences of others. Many clubs specialize in VHF and UHF operation and can be of great assistance to you. Use the resources I discuss in Chapter 3 to find these clubs.

When the bands open up or an unusual propagation event occurs, getting the word out as widely and as quickly as possible is to everyone's benefit. Sounds like a job for the Internet, doesn't it? A strong community of VHF/UHF DXers is in nearly constant communication worldwide, thanks to the Internet. My personal favorite is the set of propagation bulletin boards at dxworld.com. You can find message posting pages for all of the VHF and UHF bands, specialty pages for meteor scatter and other activities, plus links on each page to useful resources for that topic.

VHF DX awards

Because VHF DX contacts are generally not as distant or worldwide as their shortwave cousins, the VHF DX awards deal with geographic divisions on a smaller scale — *grid squares.* Grid squares are the basis for the Maidenhead Locator System, in which one grid square measures 1° latitude by 2° longitude. Each grid square is labeled by two letters (called the *field*) and two numbers (called the *square*). For example, my location near Seattle is in the CN87 grid square. Grid squares are divided even further into *subsquares,* which are denoted by two additional letters. My six-character grid square is CN87sk. (The subsquare is generally noted with lowercase letters.)

In North America, where the countries tend to be very large (except in the Caribbean), the primary VHF/UHF award program is the ARRL's VHF/UHF Century Club (VUCC). (Check the program out at www.arrl.org/awards/vucc.) The number of grid squares you need to contact to qualify for VUCC varies with the band, due to the degree of difficulty. As an example, on the lowest two bands (6-meters and 2-meters) and for contacts made using satellites, contacts on 100-grid squares are required. Along with VUCC, the ARRL's Worked All States (WAS) program has a vigorous VHF/UHF audience, as well. (For more on WAS, see the section "Getting awards," earlier in this chapter.)

In Europe, where more countries are within range of conventional VHF/UHF propagation, many of the shortwave DX awards have a VHF/UHF counterpart. Many VHF/UHF awards are based on contacting different countries, too.

What would DX-ing be without a distance record, too? On shortwave bands, with signals bouncing all the way around the world, the maximum terrestrial distance records were met long ago. In VHF/UHF, though, a lot of frontiers are still left. Al Ward, W5LUA, has put together a complete VHF/UHF/Microwave record list, which is posted at www.arrl.org/qst/worldabove/dxrecords.html. Check out the amazing list of propagation modes and distances. New records are added all the time. Maybe your call sign will be there one day. (For more on awards, see the section "Chasing Awards," later in this chapter.)

Taking Part in Radio Contests

How can you have a contest on the radio? What kind of rules are there? Are there referees? Awards? If you've never encountered a radio contest before, the concept can seem pretty puzzling.

Radio contests, or *radiosport,* are competitions between stations to make as many contacts as possible with as many different stations as possible within the time period of the contest. Time periods range from a couple of hours to a weekend. Restrictions say who can contact who and on what bands. With each contact, you must exchange specific information. Often, themes dictate which stations you are to contact, such as stations in different countries, grids, or states.

After the contest is over, participants send their logs to the contest sponsor by mail or, more commonly, by e-mail. The sponsor then performs the necessary amount of cross-checking between logs to confirm that the claimed contacts actually took place. The final score is computed and the results published in a magazine or Web page with the winners receiving certificates, plaques, or other non-monetary prizes.

What's the point of such contests? Well, for one thing, they can be a lot of fun as many stations are all on the air at once, trying for rapid-fire short contacts. In the big international contests, such as CQ World-Wide, literally thousands of stations are on the air at once on the bands from 160 through 10 meters from locations spread out all over the world! In a few hours, you can find yourself logging a WAC (Worked All Continents) and being well on your way to some of the DX awards I cover in the previous sections. Contests are a great way to make contacts for awards programs.

Contests are also a great way to exercise your station and operating ability to the limits. Are you able to crack the contest pileups? Can you copy that weak signal through the noise? Is your receiver up to the task of handling those strong signals? If you want to increase your Morse code speed, spend some time in a contest on the CW sub-bands. Just as with physical fitness, competitive activities are fun and make staying in shape a lot more fun than solitary calisthenics!

Types of contests

Contest styles run the gamut. Some contests are low-key, take-your-time events taking place on a few frequencies here and there. Other contests fill a band with hectic activity from all directions. I list some popular contests in Table 11-2. Don't worry if you don't see a contest for your favorite band, mode, or specialty — a contest exists for every taste. Use the contest calendars to find your favorites. (For more on contest calendars, see the upcoming section, "The beginning contester.")

Most contests run annually and occur on the same weekend every year. The full-weekend contests generally start at 0000Z (Friday night in the United States) and end 48 hours later on Sunday at 2359Z. You don't have to stay up for two days, but some amazing operators do! Most contests have a time limit or much shorter hours.

Table 11-2	Popular Contests
Contest Name	*Sponsor*
North American QSO Parties	National Contest Journal (www.ncjweb.com)
ARRL January VHF Sweepstakes	ARRL (www.arrl.org/contests)
ARRL DX	ARRL (www.arrl.org/contests)
CQ WPX	CQ Magazine (www.cq-amateur-radio.com)
ARRL June VHF QSO Party	ARRL (www.arrl.org/contests)
Field Day	ARRL (www.arrl.org/contests)
IARU HF Championship	IARU (www.arrl.org/contests)
Worked all Europe (WAE)	DARC (www.darc.de/referate/dx/fed.htm)
ARRL September VHF QSO Party	ARRL (www.arrl.org/contests)
CQ WW	CQ Magazine (www.cq-amateur-radio.com)
ARRL Sweepstakes	ARRL (www.arrl.org/contests)
ARRL 160-Meter Contest	ARRL (www.arrl.org/contests)
ARRL 10-Meter Contest	ARRL (www.arrl.org/contests)

Operating in contests

Don't be intimidated by the rapid-fire action that occurs during contests. Contesting is nearly unique as a sport in that the participants score by cooperating with each other. Even arch-rivals need to put each other in the log for points. The Big Guns need and want your contact. All of the participants want to talk to you.

You needn't have a huge and powerful station to enjoy contesting — most contesters have a simple setup. Besides, the most important part is the operator. By listening, knowing the rules, and having your station ready to go, you're all set.

Here's an example of a typical contest contact — in this case my own state's contest, the Washington State Salmon Run. (State contests are often referred to as *QSO Parties* to emphasize their easy-going style.) In this scenario, I'm NØAX in King County, calling CQ to solicit contacts, and you're W1AW in Connecticut, tuning around the band to find Washington stations. The information we exchange is a signal report (see Chapter 8) and my county and your state, because you're not in Washington.

> NØAX: *"CQ Salmon Run CQ Salmon Run from Norway Zero Alpha X-ray"*
>
> You: *"Whiskey One Alpha Whiskey"* (Note that you just send or say your call sign once, using standard phonetics on voice transmissions.)
>
> NØAX: *"W1AW you're five-nine in King County"*
>
> You: *"QSL, NØAX you're five-nine in Connecticut"*
>
> NØAX: *"Thanks, QRZed Salmon Run Norway Zero Alpha X-ray"*

The whole thing takes about ten seconds. Each station identifies and exchanges the required information. The "five-nine" is the required signal report signifying "loud and clear." That's an efficient contest contact and most are not much different from that. After the contact is completed, keep tuning for another station calling, *"CQ Contest."* This method of finding stations to call is known as *Search-and-Pounce* operation, or *S&P.*

What if you miss something? Maybe you just tuned in the station and the band is noisy or the signal is weak. My response to your call sounds like this:

> NØAX: *"W1AW you're five-nine in "*
>
> You: *"Sorry, please repeat your county"*
>
> NØAX: *"Kilo India Norway Golf, King County"*
>
> You: *"QSL, NØAX you're five-nine in Connecticut"* and so forth

You're probably thinking, "But I missed the county, how can the signal report be five-nine?" By convention, most contesters just use "five-nine" (or send 5NN on Morse code — the "N" represents an abbreviated "9" — a very common substitution for all contacts) because the type of report doesn't affect the score unless it's miscopied.

Contesting is no more complicated than getting your sandwich order taken at a busy deli counter during lunch hour. Contesting has a million variations, but you'll quickly recognize the basic format.

Your score consists of *QSO points* and *multipliers*. Each contact counts for one or more QSO points, sometimes depending on the mode, band, or other special consideration. Multipliers — so named because they multiply QSO points for the final score — are what make each contest an exciting treasure hunt. Depending on the theme, you may be hunting for states, grids, counties, lighthouses, islands, who knows? Read the rules carefully for how the multipliers are counted: only once, once per band, once per mode, and so on. Special *bonus points* may be awarded for working certain stations or multipliers.

The beginning contester

Start by finding out what contests are running. Many places list when contests are held or identify a contest you find on the air. Table 11-3 shows several sources, or you can enter **contest calendar** into a Web search engine. Most Web sites include the contest rules or a link to the contest sponsor's Web site.

Table 11-3	Contest Calendars
QST	**Where to Go**
ARRL Contest Calendar	www.arrl.org/contests/calendar.html
WA7BNM's Contest Calendar	www.hornucopia.com/contestcal
SM3CER's Contest Calendar	www.hornucopia.com/contestcal
ARRL "Contester's Rate Sheet"	Biweekly e-mail newsletter, free to ARRL members (www.arrl.org/contests/rate-sheet/about.html)
VK4DX's Contest Calendar	www.vk4dx.net

After you know the rules, listen to a participating station. The most important part of each contact is the information passed between stations, known as the *exchange*. For most contests, the exchange is short — a signal report and some identification such as a *serial number* (the count of contacts you made), name, location, or club membership number. By reading the rules or simply listening, you know what is required and in what order to send the information.

Logging contacts on a computer (using a special program just for contests) makes contesting easier. However, don't worry about computer logging right away. Pencil and paper is much easier to deal with as a beginner. Often, the contest sponsors have a log sheet that you can print out from a Web site containing all the required information.

If you're unsure of yourself, try "singing along" without actually transmitting. Make a "cue card" that has all of the information you need to say or send. If you think you may get flustered when the other station answers your call, listen to a few contacts and copy the information ahead of time. Serial numbers advance one at a time, so you can have all of the information before your contact.

When you're ready, give it a try. *Remember:* You don't have to be a speed demon; just be steady. Good contest operators are smooth and efficient, so just send your full call sign once. If the station answers with your call sign, log the exchange and send your information only once, even if you are using a small station. The other operator asks you to repeat yourself if some of the information is missed.

The regular contester

So, you've gotten into a few contests, but when you browse through the scores, the top scores are just out of sight! How do these guys do it? They will tell you, "There's no magic!" Winning contests just comes down to perseverance and patient practice. The tricks of the trade come with time.

Computer logging makes contesting a lot easier if you are even a modest typist. The software keeps score, maintains a *dupe sheet* (a page listing stations you already worked), shows needed multipliers, connects to DX spotting networks, and creates properly formatted logs to submit to the sponsors. I list some of the most popular software programs in Table 11-4, but just entering **contest logger** into a Web search engine turns up many different and useful programs.

Table 11-4	Popular Contest Logging Software
Name of Software	*Where Available*
CT	www.k1ea.com
TR-LOG	www.qth.com/tr
NA	datom.contesting.com
WriteLog	www.writelog.com
N3FJP Contest Loggers	www.n3fjp.com
N1MM Contest Logger	www.n1mm.com
N1MU Rover Log (for VHF contesting)	roverlog.2ub.org

If you're search-and-pouncing, use your radio's memories or alternate between the main tuning frequencies. By saving the frequencies of two or three CQ-ing stations, you can bounce back and forth between several pileups and dramatically improve your rate. Keep a list of stations by frequency (called a *band map*) to save time finding them and avoid working them a second time.

Many stations use DX spotting systems to find rare or needed stations in a contest. You can, too, but be aware that the use of any such assistance you receive usually puts you in a "multiple-operator" category. Know the rules of the contest regarding spotting assistance and be sure to submit your score and log in the proper category.

Calling CQ

To make a lot of contacts, you have to call CQ. In any contest, more stations are tuning than calling. You can turn those numbers to your advantage. Find a clear frequency (see the section "Being polite," later in this chapter) and when you're sure it's not in use, fire away. The following list offers a few examples of the appropriate transmission when calling CQ in a contest. (In this example, again assume your call sign is W1AW, and the contest is the Washington State Salmon Run. Replace these terms with those appropriate to your situation.)

- ✔ **Voice transmissions:** *"CQ Salmon Run CQ Salmon Run from Whiskey One Alpha Whiskey, Whiskey One Alpha Whiskey, Salmon Run"*

- ✔ **Morse code or digital transmissions:** *"CQ CQ TEST DE W1AW W1AW TEST"*

- ✔ **VHF/UHF transmissions:** *"CQ Salmon Run from W1AW grid FN31"*

Keep transmissions short and call at a speed at which you feel comfortable receiving a reply. Pause for two or three seconds between CQs before calling again. Other stations are tuning the band and can miss your call if you leave too much time between CQs.

When you get a stream of callers going, keep things moving steadily. Try to send the exchange the same way every time. Eliminate saying "uh" and "um." Take a breath before the exchange and say it all in one smooth sentence. As you make more contacts, your confidence steadily builds. That efficient rhythm increases your *rate* — the number of contacts per minute.

Contesting being what it is, you'll eventually encounter interference or a station that begins calling CQ on your frequency. You have two options — stick it out or move. Sometimes a simple, *"The frequency is in use, CQ contest . . ."* or

"PSE QSY"(which means "please change your frequency" in Morse code-ese) on CW does the trick. Otherwise, unless you're confident that you have a strong signal and good technique, just finding a new frequency may be more effective for you. The higher end of the bands is often less crowded and you may be able to hold a frequency longer.

Submitting a log

When you finish operating, if you want to see your score in the results, follow the instructions in the rules to submit your log. Most sponsors now accept logs via e-mail — usually as a text file. Many of the larger contests require or encourage the use of the *Cabrillo* format for logs, which is just a type of form letter that your logging software can generate. Check with the software author or contest sponsors if you're unsure. Even if you're not interested in having your score in the results, submitting your log just for the sponsor to use for checking other logs (called submitting a *check log*) is appreciated by the sponsors to improve the quality of the final scoring.

If you decide to mail in a paper log or disk, be sure all of the required information is included. Use a disk or CD mailer for protection. Many sponsors post a Logs Received Web page so that you can be sure yours was received. Don't miss the deadline for submitting logs!

Being polite

Large contests can fill up most or all of a shortwave band, particularly during voice-transmission contests, and often causes friction with non-contest operators. As in most conflicts, each side needs to engage in some give-and-take to keep the peace. If you're participating in a contest, be courteous and make reasonable accommodations for non-contesters. If you're not contesting, recognize that large competitive events are a legitimate activity and you need to be flexible in your operating expectations.

That said, how can you get along with everyone? First, be sure your signal is clean, not distorted or generating key clicks. (You might hear about such problems from stations operating near you.) A distorted signal's intelligibility is greatly reduced. A clean signal gets more callers every time and occupies less bandwidth. Keep your noise blanker and preamp off (read about these devices in the technical supplement on the *Ham Radio For Dummies* Web site or in your radio's operating manual) and use every receiver adjustment on the front panel, including the front-end attenuator.

Second, listen before you leap. Non-contest contacts are more relaxed with longer pauses, so a couple of seconds of dead air doesn't mean the frequency is clear. Asking *"Is the frequency in use?"* (or *"QRL?"* in Morse code) before calling CQ is the right thing to do whether you're in a contest or not. If a Morse code contact is ongoing, the response to your query may be just a "dit" (meaning "Yes, it's busy") if the other operator is in the middle of trying to copy an exchange from a different station. When in a contest, keep a minimum of 1.5 kHz between you and adjacent contest contacts on phone and 400 Hz on CW. Don't expect a perfectly clear channel. Contesters should tune higher in the band to find less-congested frequencies and give non-contest QSOs a little-wider margin.

Avoid major net frequencies, such as the Maritime Service Net on 14.300 MHz. Be aware of any emergency communications declarations or where regional emergency nets may meet and give those frequencies a wide berth. Calling frequencies are often busy with non-contest activity.

Learning about contesting

I've only been able to touch lightly on many important topics. After you start, many resources are available to the novice and master contester alike that can help them learn and improve. Many cost little or nothing — only the time to find and read about them.

QST and *CQ* magazines both feature contest results and articles on technique. The ARRL also publishes the *National Contest Journal* (www.ncjweb.com), which sponsors several HF contests every year and features interviews with contesters, plus articles and columns on contesting. ARRL members can receive the biweekly e-mail newsletter *The Contester's Rate Sheet* (www.arrl.org/contests/rate-sheet) without charge. The CQ-Contest e-mail mailing list is a source of many good ideas. Subscribe to it at www.contesting.com.

The best way to learn is to work with an experienced contester. Probably one or two multiple-operator stations in your region are active in the big contests. Look through the results of previous contests for their call signs. Contact the station owner and volunteer to help out — most are eager to have you on board or can help you find another team. As a rookie, expect to listen, log, or spot new multipliers — all valuable learning opportunities. When you know the ropes, you can fill in on the air more and more.

Along with the multi-operator stations, many contest clubs are around the country. Look at the club scores in the writeups and contact them. All contesters started just like you.

Chasing Awards

If the awards I mention in the section on DX-ing piqued your interest, the following sections discuss awards in greater detail. Seeking awards is one of the most fulfilling activities in the hobby of ham radio. Certificates are the usual reward and are often referred to as *wallpaper*. Indeed, some radio shacks that I've visited are often literally papered (ceilings, too) with certificates and awards. They come in all shapes, sizes, and colors. Some wallpapers are plain and other wallpapers are as colorful and as detailed as paintings or photographs. If awards sound interesting, you may be a member of the species of ham known as the *paper chaser* or *wallpaper hanger!* Believe me, a lot of them are out there!

Finding awards and special events

CQ magazine runs a column featuring novel awards every month. The *K1BV Awards Directory* lists more than 3,000 DX awards from nearly every country! Want to try for the "Tasmanian Devil" award? Contact VK7 (Tasmania's prefix) amateurs. How about the South African Relay League's "All Africa" award for contacting the six South African call areas and 25 other African countries?

Most awards have no time limit, but some span a given period, often a year. For example, the ARRL's Millenium DXCC Award was for contacts during the year 2000. Whatever your tastes and capabilities, you can find awards that suit you. You can find the awards directory and an extensive list of Web links, along with the K1BV directory at www.dxawards.com/book.html.

Along with ongoing awards programs, you can find many *special event* stations and operations. These often feature special call signs with unusual prefixes, of great interest to hams chasing the WPX award, and colorful, unusual QSL cards. Ham stations are often part of large sporting events or public festivals, such as international expositions and the Olympics. Some events are as small as a county fair, such as the Calaveras County Frog Jumping Jubilee whose station generated the novel certificate shown in Figure 11-3. The larger special event stations are well publicized and are listed on Web pages such as www.arrl.org/contests/spev.html or on the ham Internet portals I discuss in Chapter 3. Other special event stations just show up unannounced on the air, which makes finding them exciting.

Figure 11-3:
The
Calaveras
Amateur
Radio
Society
sponsored
this amusing
certificate.

Getting the contacts

Before embarking on a big adventure to achieve an obscure award, find out if it is still active by checking with the sponsors. Beware of outdated Web sites, too. Before proceeding, get a positive "go ahead" if you have the slightest question if an award program is active.

Determine whether the award requires you to submit QSL cards. Overseas sponsors may allow you to submit a simple list of contacts complying with their General Certification Rules (GCR) in place of having to submit actual cards.

When you make an eligible contact, be sure to log any information that the award may require. For example, if you're working Japanese cities, some awards may require that you get a city number or other ID. This information may or may not be printed on the other station's QSL card you receive for the contact, so you need to ask for the information during the contact itself. Grid square information is not always on QSLs, either. Be sure to ask during the contact or with a written note on the QSL you send!

If you make a contact for an award favoring a certain geographic area, ask the station if he or she would let others know you're chasing the award. That may even generate a couple more contacts right on the spot! Certainly, the request lets his or her associates know to listen for you on the band or perhaps arrange a schedule. This technique helps a lot with difficult awards or remote areas.

Applying for awards

When applying for an award, be sure to use the proper forms, addresses, and forms of payments. Make sure you follow the instructions to the letter for submitting your application. If you aren't certain, ask the sponsor. Don't send your hard-earned QSLs or money before you're convinced you know what to do.

You often see *IRCs* accepted in place of cash. An IRC is a postal system International Reply Coupon. These are sold by post offices around the world and are good for one unit of surface postage in the receiving country. Because foreign currency may not be accepted (or even legal!) in some countries, the IRC is a form of "ham dollars" that you can exchange for postage or awards. You can read all about IRCs at www.upu.int/irc/en/index.html#zone_reserve.

When you do apply for the awards, you may want to send the application by registered or certified mail, particularly if precious QSLs, cash, or IRCs are inside. Outside the developed countries, postal workers are notorious for opening any mail that may contain valuables. Make your mail look as boring and ordinary as possible. Keep envelopes thin, flat, and opaque.

QRP: Low-Power Operating

Whole books have been written about using QRP, or low-power operating (sometimes called *flea power*), and many low-power devotees are on the air. Why would you want to use low power and a weak signal instead of high power and a strong signal? For the same reason there are fly fishermen and free-climbers: skill. Putting as little as possible between yourself and the station at the other end and still making the contact takes skill. Build up a little experience and then give QRP a try.

QRP is up to 5 watts of transmitter output power on Morse code transmissions and 10 watts of peak power on voice, usually SSB. The quality of your antenna or location is not considered; just transmitter power. If you choose to turn the power down below a watt, that's called *milliwatting*.

QRP is primarily an HF activity and the majority of QRP contacts are in Morse code due to its efficiency. QRPers often hang out around their calling frequencies, shown in Table 11-5. To start QRP-ing, just tune to a clear frequency nearby and call CQ — no need to call CQ QRP unless you specifically want to contact other QRPers. Here's how it works.

Table 11-5	North American QRP Calling Frequencies	
Band (Meters)	Morse Code (MHz)	Voice (MHz)
160	1.810	1.910
80	3.560	3.985
	3.710	
40	7.040	7.285
	7.110	
30	10.106	
20	14.060	14.285
17	18.096	
15	21.060	21.385
	21.110	
12	24.906	
10	28.060	28.885
	28.110	28.385
6	50.060	50.885
2	144.060	144.285 (SSB)
		144.585 (FM)

If you're just getting started, tune in a strong signal and give them a call with your transmitter output power turned down to QRP levels (check the radio operating manual for instructions). Make sure your transmissions are clear, which allows the other station to copy your call sign easily. Some low-power stations send their call with "/QRP" tacked on to the end to indicate they're running low power. This procedure isn't really necessary during the initial call and can be confusing if your signal is weak. After all, that's four more characters for the other station to copy, isn't it?

QRPers delight in building their own equipment, and the smaller and lighter the better. You can find a lot of kits, such as the popular (and tiny) RockMite transceiver shown in Figure 11-4, and home-brew designs for the ham with good construction skills. QRPers probably build more equipment than those in any other segment of the hobby, so if you want to learn about radio electronics, you may consider joining a QRP club and one of the QRP e-mail mailing lists.

QRP-only contests and QRP categories are in nearly all of the major contests. Many awards have a special endorsement for one-way and two-way QRP. The QRP clubs themselves have a whole set of awards, of which my all-time favorite is the 1,000 Miles Per Watt award (www.qrparci.org/arciawds.html). Some stations make contact with so little power that their figures are in the millions of miles per watt!

Figure 11-4: The RockMite transceiver fits in a mints tin!

QRPers are an enthusiastic and helpful lot, always ready to act as QRP Elmers. Their clubs and magazines are full of "can-do" ham spirit. I list the QRP resources in Table 11-6, which includes the larger QRP organizations, e-mail reflectors, and magazines.

Table 11-6	QRP Operating Resources	
Resource	**Address or Source**	**Description**
QRP Amateur Radio Club, International	www.qrparci.org	*QRP Quarterly* magazine and numerous awards
American QRP Club	www.a-qrp.org	Extensive kit-building and construction resources, *Homebrewer* magazine
G-QRP Club	gqrp.com	Lots of building and operating information, *SPRAT* magazine
Adventure Radio Society	www.arsqrp.com	Emphasis is on portable operation
QRP-L e-mail reflector	listserv.lehigh.edu/lists/qrp-l	Best-known QRP e-mail reflector, includes archives for e-mail, files, and articles
QRP forum	www.eham.net/forums/QRP	Wide variety of topics
Magazine columns about QRP	*QST* "QRP Power," *Worldradio* "QRP" *CQ Magazine* "QRP"	A different technical or operating topic with every issue
QRP contests	QRP ARCI Spring and Fall QSO Parties (www.qrparci.org)	The largest low-power operating events of the year
	Adventure Radio Society "Adventure Sprints" (www.arsqrp.com)	A monthly low-power contest emphasizing portable operation
	QRP Contest Calendar by N2CQ (www.amqrp.org/contesting/contesting.html)	A comprehensive listing of QRP events throughout the year

The major QRP gatherings that occur every year are:

- **Dayton Hamvention:** This gathering is called "Four Days in May." QRPers come from all around the world to attend and enough of them fill a hotel and have their own mini-convention.

- **Pacificon:** West Coast QRP enthusiasts put on this convention in October in the San Francisco area.

- **Frederichshafen:** European QRPers get together at this giant hamfest.

Look for smaller QRP forums or mini-conventions held as part of regional ham conventions or hamfests.

Getting Digital

Operating on the digital modes (transmitting signals encoded as a stream of data, instead of speech or Morse code) is the fastest growing segment of amateur radio today. Hams are using fast computers to create some novel means of communications. Not only are hams adapting commercial technologies such as the Internet standard TCP/IP protocols, but also creating entirely new ones, such as PSK and Throb.

On HF bands, digital data transmission protocols must overcome the hostile treatment given to delicate bits and bytes by the ionosphere and atmospheric noise. The protocols (such as PSK31, PACTOR, Throb, and so on) tend to use short transmissions, with robust error detection and correction mechanisms. Limits on transmission bandwidth also contribute to low data transmission rates, but these restrictions have also stimulated hams into the creation of interesting protocols.

On VHF bands, digital data modes have fewer restrictions on bandwidth. For example, on 440 MHz and up, you can use 56k baud technology. The bands are quieter, with less fading and interference, so data speeds begin to approximate those of dial-up landline services.

Radioteletype (RTTY)

CW, or Morse code, is really the first digital mode, but the first fully automated data transmission protocol was radioteletype. Commercialized in the 1930s, RTTY, or *ritty* to hams, uses a 5-bit code known as Baudot — the origin of the word "baud" used in digital transmission today. The Baudot code sends plain text characters as 5-bit codes that use alternating patterns of two different audio frequencies known as *mark* and *space,* creating a type of modulation called *Frequency Shift Keying* or FSK.

The tones, 2125 Hz (mark) and 2295 Hz (space), fit within a normal voice's bandwidth and thus the RTTY signal can be transmitted with a regular SSB transceiver in place of a voice signal. The rate at which characters can be sent is from 60 to 100 wpm. On the receiving end, an SSB transceiver receives the transmission as an audio signal.

The text characters are turned into and recovered from the pair of mark and space tones by either a computer/sound card combination or by a standalone modulator/demodulator. Fans of the antique teleprinters that used rotating mechanical contacts and briefcase-sized tone encoding equipment keep them running and use them on the bands even today. If you get a chance to watch one at work, you'll be amazed at their mechanical complexity!

If you are using a computer to convert the RTTY signal to readable text, the decoding software will present a visual display that allows you to adjust your receiver so that the audio tones are received as the proper frequencies (2125 and 2295 Hz) for the software to decode into characters. RTTY operators sometimes send "RYRYRYRYRY . . ." which results in a repeating tone sequence of mark followed by space that allows a listening operator to tune in their signal.

Although RTTY is being supplanted by the more modern modes developed in the past decade, it is still a strong presence on the bands. Tune through the digital signals above the CW stations and you hear lots of two-tone signals "diddling" to each other. A sizeable community of RTTY DXers and several major award programs have RTTY endorsements. DX-peditions often include RTTY in their operating plan, as well. Table 11-7 lists several online resources for beginning RTTY operators.

Table 11-7	RTTY Resources
Resource	*Description*
AA5AU's RTTY Page (`www.aa5au.com/rtty`)	Web site with lots of tutorial information, links to RTTY programs, troubleshooting, RTTY contesting
RTTY e-mail reflector (subscribe at `lists.contesting.com/mailman/listinfo/rtty`)	International membership e-mail group
MMTTY software site (`www.qsl.net/mmhamsoft/mmtty`)	Most popular RTTY software for computers
DJ3NG's RTTY Contesting Site (`www.rtty-contest-scene.com/index1.html`)	RTTY contest calendars, results, and upcoming events

TOR modes — AMTOR, PACTOR

TOR is an acronym for *Teleprinting Over Radio,* which means that TOR systems send text characters. A user of RTTY quickly discovers that the fading and distortion common on HF can do serious damage to characters sent using the Baudot code. The TOR systems include data organization and error-correction mechanisms to overcome these limitations.

AMTOR was the first TOR system for amateurs, adapted from a commercial technology called SITOR in the early 1980s. Like RTTY, AMTOR uses frequency-shift keying to generate the on-the-air audio. Characters are 7 bits long (Baudot uses 5) and sent in short blocks with pauses between them to allow other stations to respond. This system makes AMTOR transmissions sound like regularly-spaced chirps on the air. Two stations in contact using AMTOR chirp alternately at each other as they exchange the blocks of characters.

AMTOR has two modes, depending on whether a station is calling CQ or in contact with another station. When calling, because no receiver reports errors, blocks of five characters are sent twice. After two stations establish contact, the transmitting station transmits groups of three characters, pausing for a "received OK" chirp from the receiving station. AMTOR has proven to be reliable, but is fairly slow and cannot send regular data.

PACTOR addresses some of AMTOR's flaws with an improved data structure and better error management mechanics. Less overhead is associated with transmissions using PACTOR and the protocol adjusts its speed based on conditions. PACTOR II is an improved version that is backwards-compatible with PACTOR, but has the ability to use a more advanced modulation technique called *Phase Shift Keying* (PSK) to increase the data rate. The management of the transmission and reception process is also much improved over PACTOR. PACTOR III is the most recent release of this technology. All versions of PACTOR are proprietary designs and only available in equipment available from the manufacturer, Special Communications Systems.

Other similar data modes include the proprietary CLOVER, CLOVER II, and CLOVER 2000 protocols developed by HAL Communications. These modes use increasingly intelligent management of the way the data transmission and reception protocols react to band conditions. The transmitted waveform shapes and frequencies are carefully managed to keep the signal within a 500 Hz bandwidth and decrease errors caused by HF propagation.

You can find a discussion of the TOR modes mixed in with a discussion of all of the other digital modes, so the best resources for AMTOR and PACTOR are on Web pages that cover many different modes. The ARRL digital data Web page at www.arrl.org/tis/info/digital.html has links to numerous useful sites.

Packet

Packet is short for packet radio, a radio-based networking system based on the commercial X.25 data transfer protocol. Developed by the Tucson Amateur Packet Radio group, or TAPR, packet's ability to send error-corrected data over VHF links enabled a number of novel new data systems to be created for hams. Table 11-8 lists three excellent online resources for packet radio users.

Table 11-8	Packet Resources
Resource	*Description*
TAPR Packet Radio page (`www.tapr.org/tapr/html/pktf.html`)	Encyclopedic collection of links and tutorials
ARRL Introduction to Packet (`www.arrl.org/tis/info/pdf/49244.pdf`)	Article explains packet radio operation
Packet newsgroup (`rec.radio.amateur.digital.misc`)	Discussions include packet radio topics

With packet, ordinary VHF/UHF FM transceivers transfer the data as audio tones. An external modem, called a Terminal Node Controller (TNC), Multimode Communications Processor (MCP), or sometimes a Multiple Protocol Converter (MPC), provides the interface between the radio and a computer or terminal. Data is sent at either 1200 or 9600 baud as packets of variable length, up to about 1,000 bytes. The protocol that controls packet construction, transmission control, and error correction is called *AX.25,* for Amateur X.25. Packet systems can also use the TCP/IP protocol and a number of packet systems use it.

Like wired networks, packet systems are connected together in many different ways. Packet networks extend across most of the United States and Europe, with gateways to the Internet in several places. A packet controller is called a *node.* The connection between nodes is a *link.* Connecting to a remote node by using an intermediate node to relay packets is called *digipeating.* A node that does nothing but relay packets is a *digipeater.* A node that makes a connection between two different packet networks or between a packet network and the Internet is called a *gateway.*

Packet radio, using its 1200- or 9600-baud protocol, does not attempt to compete with the higher speeds of WLAN technology. A number of groups are now beginning to use WLAN hardware on the amateur bands (see the section,

"Amateur WLAN and high-speed data"). Packet remains in use around the United States for now — particularly as an effective mode of emergency communications — but look for it to gradually be replaced by more modern technology.

Bulletin boards

The most common use of packet is to support a packet bulletin board system (PBBS). A PBBS typically has users from a local area that can contact the PBBS via a VHF band either directly or by digipeating. The PBBS is operated just the same as a regular telephone dial-up BBS. In fact, some BBSs support both types of access.

The PBBS offers file transfer and storage, bulletins, network relay, mailboxes, and other useful functions. Although the Internet has reduced the importance of the PBBS in favor of the Web, packet systems are still quite common and are frequently deployed to support emergency communications.

Packet Cluster

The DX spotting networks now so prevalent on the Internet started as local groups of DXers using packet radio to relay information about DX stations. Packet Cluster is a special form of bulletin board software designed by Dick Newell AK1A to act as an information service for DXers to distribute spots of DX stations. These bulletin boards are referred to as *clusters* because the group was configured as one central information server station surrounded by client stations that connected to the server by a VHF radio link using the packet radio AX.25 protocol. Along with DX spots, clusters now handle text messages, data files, weather, and solar activity data. Packet Cluster has been adapted for commercial and public safety use worldwide.

The clusters were originally limited to local groups, but quickly grew to regional networks that now may have hundreds of users connected at any given time. The radio cluster networks are also connected to the Internet spotting networks through gateways, so information is collected locally on VHF, passed through the packet network, exchanged with a network on a different continent via the Internet, and then re-sent via VHF.

HF packet

Although it is poorly suited for the conditions encountered on HF, you can make packet work below 30 MHz. The main difficulty is that AX.25 packets are too long for the rapidly changing and noisy HF channels and so an excessive number of packets are rejected due to errors. Retransmission is very inefficient and so HF packet's *throughput* is not very high, leading to it being rarely used on the HF bands.

PSK modes

The most exciting new digital mode for use in the HF spectrum is PSK31. Peter Martinex G3PLX invented the new mode and developed a complete package of Windows-based software to support it. He generously placed his creation in the ham radio public domain and hams are adopting it like wildfire. If you want to find out more about using PSK31, Table 11-9 contains several good reference Web links.

PSK stands for *Phase Shift Keying* and the 31 represents the miniscule 31.25 Hz occupied by a PSK31 signal. It also uses a new coding system for characters, called Varicode, that has a different number of bits for different characters — not unlike Morse code. Instead of turning a carrier on and off to transmit the code, a continuous tone is transmitted that signifies the bits of the code by shifting its timing relationship (known as *phase*) with a reference signal. A receiver syncs up with the transmitter and decodes even very noisy signals, because the receiver knows when to look for the phase changes.

When you tune across a PSK31 signal, the signal doesn't chirp, but rather warbles in a very characteristic way. Because the human ear has difficulty distinguishing between pitches so close together, tuning by ear alone is difficult. This difficulty brought about a new visual tuning technique, not too different than that used for RTTY.

PSK31 is very tolerant of the noise and other disturbances on HF bands. In fact, you can obtain nearly solid copy with signals barely stronger than the noise itself. Figure 11-5 shows the DigiPan software display of several signals, some quite weak. The lighter streaks represent signals and each horizontal line represents a new sampling of the receiver's output. New signals appear at the top and slowly drift downwards, thus earning the display the "waterfall" name.

Table 11-9	PSK Resources
Resource	*Description*
Tutorial on PSK31 (www.arrl.org/ tis/info/HTML/psk31/index. html)	Introduces how PSK31 works and how to operate on the mode
PSK31 Home Page (aintel.bi.ehu. es/psk31.html)	The latest updates on the mode and software to use it
DigiPan Home Page (www.digipan. net)	Free software to operate PSK31 with the waterfall tuning display
PSK Mailing List (aintel.bi.ehu. es/psk31.html)	E-mail reflector for discussion of PSK31 and the latest variants

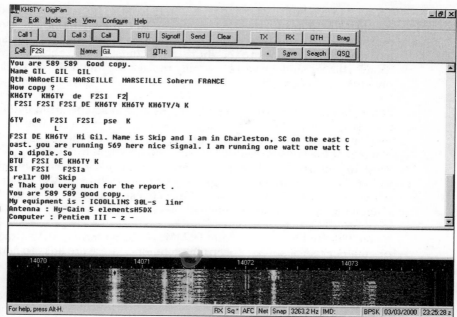

Figure 11-5:
A PSK31
QSO using
the waterfall
display of
DigiPan.

Because the bandwidth of PSK31 is so narrow, finding other PSK31 stations on the air requires a pretty good idea of where they are. The most common frequencies are 3580.150, 7080.15, 10142.150, 14070.150, 18100.150, and 21080.150 MHz.

Since PSK31's introduction, a number of enhancements have been made to the original protocol, including a variant called PSK63 that adds some features and quality improvements at the expense of wider bandwidth. PSK31 contests are held everywhere, and a growing community of PSK31/63 users can be found around the world.

Amateur WLAN and high-speed data

The widespread adoption of wireless LAN technology (WiFi, 802.11b, and 802.11g) by computer users has brought the same technology within reach of amateurs, as well. Hams share the 2.4 GHz band with unlicensed LAN devices, but without the power restrictions of consumer equipment.

Ham experimenters are in the process of adapting commercial WLAN protocols to Amateur Radio, with higher transmitter power and larger antennas than commercial technology to support a set of developmental protocols known as High Speed Multimedia Radio (HSMM). The network they envision is called the Hinternet (for Ham Internet). With the higher power levels permitted for amateurs, their intent is to create long-distance data links that run at equivalent data rates to the short-range commercial technology available today. This technology will eventually replace the much slower packet radio networks in place today.

Some experimental work is being done using spread spectrum modems on the UHF and microwave ham bands. At present, this is a relatively small community of experimenters that is attempting to extend the range of commercial technology on amateur frequencies. Momentum is beginning to build for these modes and the coming years are likely to see an upsurge in amateur use of these technologies. An amateur satellite (dubbed PANSAT) that would use spread spectrum modes is in the design stages. TAPR hosts the largest community of spread spectrum enthusiasts.

If you'd like to know more about HSMM, TAPR, Hinternet, RLAN (Radio LAN), or amateur use of spread spectrum techniques, here are links to active groups:

- **HSMM Web site:** www.arrl.org/hsmm
- **TAPR Web site:** www.tapr.org
- **KB9MWR Web site:** www.qsl.net/kb9mwr/projects/wireless/plan.html
- **Green Bay Professional Packet Radio:** www.gbppr.org

Amateur radio is also beginning to see the first digital voice links: Icom, one of the largest radio manufacturers, has introduced its D-Star system of radios, repeaters, and data link equipment. D-Star is an open standard and amateurs and other groups are being encouraged to develop interfaces to this equipment. The system uses a high-speed data link that can also be used to transfer data. You can find out more about D-Star at www.icomamerica.com/amateur/dstar. If history is any guide, other high-speed data technologies will soon follow. This is an exciting time for digital innovation in ham radio.

APRS

The Automatic Position Reporting System (APRS) is an amateur invention that marries GPS positioning and packet radio, and was developed by Bob Bruninga WB4APR. Stations using APRS connect the location data from GPS receivers to a 2-meter radio and transmit "I am here" data packets. The packets are either received directly by other hams or by a packet radio digipeater.

If they are received by the digipeater, they may also be received by a gateway station that relays the call sign and position information to an APRS server accessed through the Internet.

After the information is received by a server, it's accessed by using a Web browser or an APRS viewing program. Figure 11-6 shows a map with the location of WA1LOU-8. ("WA1LOU" is a call sign. The -8 is a *Secondary Station ID* or *SSID* so that the call sign is used for several different purposes with different SSIDs.) You can see that WA1LOU is in mid-Connecticut, 7.9 miles northeast of Waterbury.You can also zoom in on WA1LOU's location. If he changes his position, this change is updated on the maps at the rate he decides to have his APRS system broadcast the information.

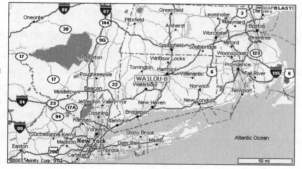

Figure 11-6: WA1LOU's position reported by APRS.

If you have a GPS receiver with an NMEA 0183 data output port (NMEA is the National Marine Electronics Association) and a 2-meter rig and packet radio TNC (see the discussion on Packet earlier in this chapter), you're ready to participate! (Kenwood even makes APRS-ready 2-meter radios, the TH-D7A and TH-D700, that are connected directly to a GPS receiver.) The most common frequency for APRS is 144.39 MHz, although you can use 145.01 and 145.79 MHz. You can find a group of HF APRS users using LSB transmission on 10.151 MHz. (The actual tones are below the carrier frequency of 10.151 MHz and so fall inside the 30-meter band.)

You can do a lot more with APRS than just report location. Popular mapping software offers interfaces so that you can have street-level maps linked to your position in real time by ham radio! Race organizers use APRS to keep track of far-flung competitors. You can also add weather conditions to APRS data to contribute to a real-time automated weather tracking network. To find out more, including detailed instructions about configuring equipment, the resources listed in Table 11-10 are good starting points.

Table 11-10	APRS Resources
Resource	**Description**
"Position Reporting with APRS" (www.arrl.org/tis/info/HTML/aprs/pos-reporting.html)	Primer on APRS technology and use
APRS Tracks, Maps and Mobiles, a book on APRS by WA1LOU	Discover how to track anything that moves; available from ham radio book vendors
APRS Home Page (web.usna.navy.mil/~bruninga/aprs.html)	Bob Bruninga WB4APR's Web page describes the current state of the technology and has many useful articles and links
TAPR APRS group (www.tapr.org/tapr/html/gpsf.html)	Extensive information on APRS equipment, tutorials, mailing lists, and software
N1BQ's APRS search page (www.wulfden.org/aprsquery.shtml)	Excellent APRS map server interface; try entering your zip code into the APRS activity or weather station search windows

Operating on the digital modes

You can gain access to the digital modes at very low cost by building or buying a simple interface between an HF or VHF radio and a computer with a sound card. This setup allows you to connect your radio and computer together (the topic of Chapter 15). To use the proprietary PACTOR II or CLOVER protocols, you need an external MCP (Multimode Communications Processor) that acts as a modulator/demodulator for several different data transmission protocols. A long list of suitable software for both sound cards and MCPs is at home.wanadoo.nl/nl9222/digisoft.htm. To pick up digital modes quickly, the ARRL has an excellent online course, HF Digital Communications EC-005 (www.arrl.org/cce/courses.html#ec005).

You find digital data at the upper end of the CW (Morse code) segments of the HF bands. The signals are mixed in with CW stations and the different modes generally get along pretty well. RTTY calling frequencies are on several bands: 3.590, 7.040, 14.080, 21.080, 28.080, and 50.7 MHz. You can find the burgeoning PSK community in the vicinity of 3.580, 7.080, and 14.070 MHz.

The overall band plans at www.arrl.org/FandES/field/regulations/bandplan.html help you locate digital mode watering holes.

Operating via Satellites

Non-hams are usually pretty surprised when you tell them about ham radio satellites. The first amateur satellite, OSCAR-1 (which stood for Orbiting Satellite Carrying Amateur Radio), was built by American hams and went into orbit in 1961, just a couple of years after the Russians launched the Sputnik satellite, igniting the space race. Today, hams have quite a number of satellites with missions ranging from digital mailboxes to repeaters to scientific experiments. The latest to go up was built by a group of Saudi Arabian students and hams; SO-50 or Saudisat 1C. It carries several experiments and a 2-meter FM repeater.

Satellite basics

Most amateur satellites are located in near-circular *Low Earth Orbit*, or *LEO*, circling the earth a number of times each day. A few have non-circular "Molniya" orbits that take them high above the earth where they are visible for hours at a time. (Molniya is "lightning" in Russian and is the name given to their fleet of communications satellites that travel in elliptical orbits.)

For practical and regulatory reasons, satellite transmissions are restricted to the bands on 10-meters; on the 2-meter, 70-cm; and microwave bands at 1296 MHz and higher. The ionosphere usually does not pass signals at lower frequencies and satellite antennas need to be small, requiring shorter wavelength.

The input frequencies are called the *uplink* and the output frequencies are called the *downlink*. The numbers that describe a satellite's orbit (and allow software to determine where it is) are called the *orbital* or *Keplerian elements*. These pieces of information allow you to operate using a satellite!

You find three common types of satellites.

- **Transponder:** These satellites listen on a range of frequencies on one band, translate those signals to a different band, and then retransmit them in real time.

- **Repeater:** These satellites act just like terrestrial repeaters, listening and receiving on a specific pair of channels. Satellite repeaters are *crossband*, meaning their input and output frequencies are on different bands.

✔ **Digital:** Digital satellites can act as bulletin boards (BBS) or as *store-and-forward* systems. You can access both types of digital satellites using regular packet radio protocols and equipment. The International Space Station (ISS) and Space Shuttle (STS) both have digital BBS systems available to hams on the ground. The ISS also has an APRS digipeater onboard! Store-and-forward satellites act as message gateways, accepting messages and downloading them to a few control stations around the world. The control stations also pass messages back up to the satellites that are downloaded by ground-based users. Digital satellites are very useful to hams at sea or in remote locations.

Accessing the satellites

The best place to go to find out which satellites are active and in what mode is the AMSAT home page (www.amsat.org). Click the <u>Satellite Frequencies and Status</u> link to get the complete set of information on what each satellite does and its current operational status.

To access satellites, you also need a satellite tracking program. Several free or shareware trackers are available, such as STSPLUS, which is available for a $15 donation from www.dransom.com/stsplus.html. AMSAT also provides several tracking and satellite operation programs. After you have the tracking software, obtain the Keplerian elements for the satellite you're seeking from the AMSAT site at www.amsat.org/amsat/keps/menu.html. Enter them into your software, and make sure that your computer's time and date are correct.

A complete set of instructions on using the satellites is beyond the scope of *Ham Radio For Dummies,* but here is a short example of how to connect to the packet station aboard the ISS and receive a nice certificate!

You need a 2-meter FM radio with 10 to 25 watts of output. An omnidirectional antenna works and a small beam is even better. You also need a packet radio TNC, such as a Kantronics KPC-3 or the equivalent. You need a computer running a terminal program, such as Hyperterm or Procomm, logging information both to and from the TNC.

1. **Configure the TNC parameters as follows:**

AUTOCR	OFF
LFADD	OFF
MAXFRAME	4
MCON	ON
MCOM	ON (which allows you to see packets from other stations)

MONITOR	ON (ditto)
PACLEN	72 (keeps packets short for fewer errors)
RETRY	8-10
TIME STAMP	ON (logs your data to disk)

2. **Set your transceiver to receive on 145.800 MHz and transmit on 145.990 MHz.**

3. **Find the right time.**

Use the tracking software to determine when the ISS has a long high-angle pass near your location with an elevation of 40 degrees or more to give you a clear path and a long view. When the time comes, start listening to the receiver audio.

4. **Get the signal.**

When you hear a signal, your TNC decodes the packet messages. If the ISS is already connected to another station, you probably see *Index* packets, such as

```
"RSOISS-1>NOAX <<I2>>:To : NOAX"
```

if I (NØAX) am logged onto the ISS system.

Only one station can be logged on at a time, so you need to be sure you aren't hearing packets from anyone else before you begin transmitting. If you do hear packets, watch your screen and wait for this message:

```
RSOISS-1>CQ/V [12/30/02 04:20:59]: <<UI>>: - Logged off
             "NOW YOU MAY BEGIN CALLING ISS, C RSOISS"
```

5. **Respond to the ISS.**

Tell your TNC to connect to the ISS by typing **C RS0ISS-1**. If you are successful, you receive a message that looks like this:

```
30-Jun-02 09:03:43 CONNECTED to RSOISS-1
           Logged on to RSOISS's Personal Message System on
           board the International Space Station
```

If you are unsuccessful or the ISS system is busy, another pass should occur shortly — the ISS is up there 24 hours a day. Keep trying. Sooner or later you'll get through!

6. **Success! Begin celebrating!**

After you calm down, politely log off by typing **BYE** or by using Command mode of the TNC to disconnect.

You can find a complete discussion of accessing the ISS at `www.marexmg.org/fileshtml/unprotopage.html`. I highly recommend you read this page before calling the ISS.

Seeing Things — Image Transmissions

So far, all of the ham transmissions I covered have either been voice, data, or codes. Don't hams care about pictures and graphics? They do! With the increasing availability of excellent cameras and computer software, getting on one of the amateur image modes, such as slow-scan or fast-scan television and facsimile, is never easier. The next few sections discuss these image modes. Figure 11-7 shows examples of images sent on each mode.

Figure 11-7: Pictures typical of those sent by amateur radio image modes.

Slow-scan television (SSTV) and Facsimile (Fax)

Pioneered back in the 1960s, slow-scan has re-emerged in the last ten years as a popular image mode. You can find slow-scan primarily on HF where SSB voice transmission is the norm. The name comes from the fact that transmitting the picture over a narrow channel made for voice transmissions takes several seconds.

You can send slow-scan pictures using a commercial *scan-converter* that converts the regular camera video to slow-scan formats and an HF transceiver. SSTV enthusiasts are rapidly eliminating the standalone converters in favor of a computer with a video camera interface and a sound card. Software is used both to convert the camera video to data files and to encode and decode the data files as audio that can be transmitted and received with a voice SSB transceiver.

You can usually hear slow-scan signals in the vicinity of 14.230 and 21.340 MHz using USB transmissions. Depending on the picture format you use, the signals make a "loopy loopy loopy" sound with the video and color information being encoded as different audio tones. Other tones mark the edges of the picture frame. Each "loop" is one line of picture. Data file transmissions may sound more like multiple tones mixed together.

Facsimile-over-radio was at one time a very widely used method of obtaining weather information from land-based and satellite stations. Transmissions of Group I and Group II facsimile standard signals could be successfully made via ham radio transceivers, as well. The use of standalone fax machines and modems, never very widespread, is rapidly disappearing in favor of transmitting the images as data files, just as with SSTV. Regular fax images can still be received from orbiting weather satellites, however.

The software packages that transmit data files as SSTV images often make no distinction between data captured from video and data received from a fax service. You will find many good links to detailed information about slow-scan TV and facsimile transmission at these Web addresses:

- www.ac6v.com/opmodes.htm#SS
- www.arrl.org/tis/info/sstv.html
- www.arrl.org/tis/info/wefax.html

Fast-scan television

You can also send full-motion video, just like regular broadcasters, with fast-scan video transmissions. Fast-scan uses the same video standards as broadcast and consumer video, so you don't need to buy special conversion equipment. This mode is usually referred to as *ATV* or Amateur Television. It is most popular in metropolitan and suburban areas where transmission distances are relatively short. ATV even has its own repeaters!

ATV transmissions are restricted to the 70-cm band and higher frequencies because of their wide bandwidth — up to 6 MHz. You won't be able to use your regular 70-cm transmitter to handle that bandwidth, so you must construct or purchase a transmitter designed specifically for ATV. The transmitters are designed to accept a regular video camera signal, so little extra equipment, except a good antenna, is required to use ATV.

Because ATV transmissions use the same video transmission format as TV broadcasters, regular television receivers are used as receivers, with a frequency converter to transfer the ham band ATV signals to one of the higher UHF TV channels where they are received just like any other TV signal.

Part IV
Building and Operating a Station That Works

The 5th Wave   By Rich Tennant

CRUDE HAM SHACK

In this part . . .

To take part in all the ham radio fun stuff, you have to assemble your own ham radio station. A station can be pretty simple, but the more you know, the easier setting yours up is. You start by figuring out what you want to accomplish with your ham radio and what your resources are. Then you can choose your radios, antennas, and accessories.

After you get your station together, you discover how to operate the station and keep simple records. You develop the savvy to get on the air with a good signal and a deft hand at the controls. (I wish this book had been around when I was just getting started!)

Chapter 12

Getting on the Air

*O*ne of the most common questions an Elmer (mentor) hears (aside from "What's the answer to number 22?") is "What kind of radio should I buy?" An honest Elmer always replies, "That depends!" Which antenna to buy depends on a whole lot of things, too!

Even a casual stroll through the ads in *CQ* or *QST* turns up page after page of colorful photos, with digits winking, lights blinking, and meter needles jumping. Antennas are even more numerous, with elements sticking out every which way, doodads dripping off of them, and all manner of claims made about performance. Then you have to sort through nearly an infinite number of accessories and software packages. The decision can be overwhelming! How do you choose between them?

Setting Goals for Your Station

Don't tell anybody, but you're about to embark on a journey called "system design." You may think making decisions is impossible, but all you have to do is a little thinking up front.

Goals and personal resources

You can find a lot of different activities in ham radio — casual, competitive, technical, public service — which I cover in Part III. While participating in many of them using the same equipment is possible, before you begin acquiring additional equipment is the time to consider these questions:

✔ What attracted you to ham radio the very first time?

✔ After reading about the different operating activities and modes available, can you pick two or three that really pique your interest?

✔ If you know and admire a ham, does he or she do something that you want to do?

✔ Are you most attracted to the shortwaves or to the VHF and UHF bands?

✔ What sounds most intriguing — the new digital modes, chatting by voice, or mastering Morse code?

✔ Will you operate from home, while mobile, or portable? (Or all three?)

✔ Do you intend to participate mainly for enjoyment or for a specific purpose, such as emergency or travel communications?

All these considerations color your choice of equipment. Knowing your ham radio resources is also important:

✔ What's your budget for getting on the air?

✔ How much space do you have available for your shack?

✔ How much space do you have for antennas?

Now comes the fun part — shopping and choosing!

To get an idea of what products are available, the advertisements in recent copies of *QST, CQ,* and *WorldRadio* magazines show models that are recently introduced. If you have a license, no doubt you receive a copy of a catalog from Ham Radio Outlet (HRO) or Amateur Electronic Supply (AES), two of the large, nationwide equipment distributors. Perhaps MFJ Enterprises sent you a catalog with its extensive line of accessories. If you have a local radio store, make a visit and browse through the catalogs and product brochures. Inquire about upcoming sales or promotions. The goal is to gather a wide variety of information.

Operating from home

A home station is a semi-permanent installation. Along with the radio equipment, you need a little furniture and space to put it in. Operating using voice modes means speaking out loud and probably listening on a speaker. Choosing an appropriate location for your station can minimize impact on other family members. For example, a basement shack should not be right under a bedroom. Garage and attic shacks have wide temperature swings. All in all, a spare bedroom or dry basement area is about the best place. Try to set up the radio shack somewhere that is not disruptive and put a good pair of headphones on your shopping list.

Because most hams operate with external antennas, plan appropriate ways of getting feedlines to them. What's going to hold the antennas up? Larger structures, such as rotatable beams on masts or towers, may need permits or approvals.

A big part of the amateur service is being available in emergencies. Because you may lose power when it's needed most, consider how you might operate your station with the AC power off. A radio that runs off of 12 volts can run from a car battery for a while. All your computing gear and accessories also need power. If you have a generator, consider how you can power your station, if necessary.

Mobile operation

The small all-band, multi-mode radios available today can put HF, VHF, and UHF bands at your fingertips, so you need to consider those possibilities. Having an efficient HF mobile antenna is harder, so skew your budget for mobile antennas towards the HF antenna.

Driving your station creates its own set of unique considerations. Because vehicles come in so many styles, you need to make a custom installation. Leave some budget for automotive fixtures and wiring. You may find spending a few dollars to have a professional shop make recommendations about power wiring and safety, in particular, prudent.

You can get lots of good ideas by reviewing articles and Web sites featuring mobile operation. I like K2BJ's Web site (www.k2bj.com), and a number of links to other good sites is at www.arrl.org/tis/info/HF-Mobile.html.

Portable operation

With many levels of portable — from minimalist backpacking to parking the RV — you need to consider being able to carry or pack your entire radio station, including your power source. Start by assigning yourself a total weight budget. Get creative on selecting antennas and accessories to maximize your options for the radio and power.

Some amazingly small radios are available. These radios aren't always the easiest to operate, however. If you're just starting out, you may want to pass up the smaller radio in favor of a rig that's easier to operate and has more features in order to learn more about operating. With more experience, you'll know what features you can do without.

If your station or operating time is limited, concentrate on a single band as you get started. On HF, 14, 17, and 21 MHz are favorites with the low-power and portable operators. These frequencies are open for a large portion of the day and the antennas are small enough to make carrying easy. If you like nighttime operating, 7 and 10 MHz are best. On VHF, 50 and 144 MHz are preferred. Plenty of operators are on those bands which feature interesting propagation.

Operating a hand-held radio

Regardless of what other pursuits you choose in ham radio, you probably want to have a handheld VHF/UHF radio. They're just so darn handy! The hand-held radio can keep you in touch with local family and friends. They're very useful on club and personal outings. Many hand-held radios also feature an extended receive range that may include commercial broadcast stations or police and fire department bands.

If you're buying your first hand-held radio, get a simple, single-band model. You can make a much more informed decision later if you decide to upgrade to a multi-band model with all the bells and whistles. Simple radios are also easy to operate. No new radio you can buy will be missing any significant feature.

If you buy an older used VHF/UHF radio — either hand-held or mobile — be sure that it has sub-audible tone features built-in. These radios often required accessory boards or modifications to include the tones. These days, sub-audible tone is required, and finding the parts to modify an older rig is hard.

Accessories can extend the life and usefulness of a portable radio, such as the following:

- ✔ The flexible rubber duck antenna supplied with hand-held radios is well suited for portable use, but isn't as efficient as a full-sized metal antenna. An external antenna greatly extends the range of a hand-held radio while at home.

- ✔ Use a high-quality, low-loss feedline for cables of more than a couple of dozen feet. (See the section titled "Feedline and connectors," later in this chapter, for more information on feedlines.)

- ✔ A speaker-mike combination allows you to control the radio without having to hold it up to your face.

- ✔ A case or jacket protects the radio against the rough-and-tumble nature of portable use.

- ✔ Spare batteries are a must! If you have a rechargeable battery, be sure to have a spare and keep it charged. A drop-in charger recharges batteries faster than the supplied wall-transformer model. If the manufacturer offers one, a battery pack that accepts ordinary AA-cells is good to have, especially in emergencies when you may not be able to use a charger.

Regardless of what kind of radio you have, be sure to keep a record of model and serial numbers. Engrave your name and driver's license number on the case in an out-of-the-way location. Mobile and portable radios can be lost or stolen (fate forefend!). Even larger radios sometimes get pressed into service on portable expeditions. Protect your investment against theft and loss! Check your homeowner's and auto insurance for coverage of radio equipment.

Allocating resources

When you start assembling a station, you have a range of items to obtain. Not only do you have the radio itself, but antennas, accessories, cables, and power sources. Table 12-1 shows some estimates of relative costs based on the type of station you're setting up. If you pick a radio first, the remaining four columns give you a rough idea of how much you should plan on spending to complete the station. These figures are approximate, but can get you started. I assume all the gear is purchased new.

Table 12-1	Relative Expense Breakdowns				
	Radio and Power Supply or Batteries	*Antennas*	*Accessories*	*Cables and Connectors*	*Total Cost Relative to Basic HF Base*
Handheld VHF/UHF	75%	Incl.	25%	Incl.	0.3
Mobile VHF/UHF	75%	20%	Not req'd	5%	0.5
All-Mode VHF/UHF	50%	30%	5%	15%	1.0
Portable HF	75%	10%	10%	5%	0.5
Mobile HF	60%	25%	10%	5%	0.7
Basic HF Base	50%	25%	15%	10%	1.0
Full-Featured HF	75%	15%	10%	5%	2.0

Choosing a Radio

As you can see in Table 12-1, regardless of what kind of station you plan on assembling, a new radio consumes at least half of your budget, which is only appropriate because the radio is the fundamental piece of equipment in ham radio. You interact with the radio more than any other equipment and a poor performer is hard to compensate for.

HF or shortwave radios

HF radios for the home station fall into three basic categories: Basic, Journeyman, and High-Performance. Modern radios all have perfectly useable receive and transmit performance. The differences are found as improved performance in several key areas: ability to receive in the presence of strong-signals, receive signal filtering and filter control capabilities, coverage of one or more VHF/UHF bands, operating amenities such as sub-receivers, number of memories, and built-in antenna tuners to name a few.

- **Basic:** This radio includes a simplified set of controls with basic receiver filter and signal adjustments. Controls may be fixed-value with on and off settings. It also has limited displays and metering, connects to a single antenna, and has little support for external accessories. A basic radio is good for a beginning ham and makes a great second or portable radio later. A computer interface may be available.

- **Journeyman:** This radio includes all the necessary receive and transmit adjustments; most with front panel controls. It has an expanded set of memory, display, and metering functions. You can find models that have additional bands and support for digital data operations. Internal antenna tuners are common as are connections for external equipment, such as transverters and band-switching equipment. A computer interface is available for control by a computer.

- **High-Performance:** Extensive set of receive and transmit controls are available on the front panel or are configurable under a menu system. A state-of-the-art receiver and sub-receiver are included along with complete interfaces for digital data and computer control. Internal antenna tuners are standard and some antenna switching is usually provided. Complete displays and metering, computer-style displays are popular.

Table 12-2 lists some of the more popular candidates in each class. Just the base models are listed. You may find various suffixes tacked on to indicate enhancements. For example, the IC-746PRO is an enhanced version of the IC-746, the Elecraft K2 comes in 5-watt and 100-watt output versions, and the FT-1000 family includes several different variations.

Table 12-2	Home Station HF Transceivers		
	Basic	*Journeyman*	*High-Performance*
Alinco	DX-77		
Elecraft		K2	
Icom	IC-718	IC-746 (incl. 50/ 144 MHz)	IC-756 (incl. 6-meters)

	Basic	*Journeyman*	*High-Performance*
Kenwood	TS-570 (S-model incl. 6-meters)	TS-870, TS-2000 (incl. 50/144/440/ 1200 MHz)	
SGC	SG-2020 (20 watts output)		
Ten-Tec	Argonaut (20 watts output)	Jupiter	Orion
Vertex Standard (Yaesu)	FT-600	FT-840, FT-920 (incl. 50 MHz), FT-847 (incl. 50/ 144/440 MHz)	FT-1000

Choosing a filter

In order to keep nearby signals from interfering with the desired signal, a receiver uses filters. These filters must pass the desired signal while *attenuating* (reducing the strength of) unwanted signals just a few hundred Hz away. For many years, the only filter components able to accomplish this feat were quartz crystals, so this type of filter is referred to as a *crystal filter.* A *mechanical filter,* similar to crystals, uses vibrating discs. Receiving filters are applied to the radio signal before it is converted to audio. These filters have a fixed *bandwidth* (the range of frequencies they can pass, measured in Hz) and cannot be adjusted. Find more information on filters on the *Ham Radio For Dummies* Web site.

Fixed-width filters are available with several bandwidths. A radio is always shipped with at least one SSB (single sideband) filter installed (HF radios) or an FM filter (VHF hand-helds and mobiles). The standard filter bandwidth for HF SSB operation is 2.4 kHz. Filters with widths of 1.5 to 2.0 kHz are available for operating under crowded conditions with some loss of fidelity. For Morse code and digital data, the standard filter is 500 Hz wide and is a good option to select. Narrower filters down to 250 Hz are available. The most common filter option to buy is the 500 Hz CW filter, followed by a narrow SSB filter.

Some Journeyman and High Performance-class receivers allow filters to *cascade* (in other words, the filters are used one following the other) at more than one intermediate stage. If you can afford the extra expense, purchasing the extra filters can often make a significant difference in receiving ability, particularly on a crowded band. To see how cascading works, check out the *Ham Radio For Dummies* Web site.

Digital Signal Processing (DSP, the P can also stand for *processor*) refers to a microprocessor in the radio running special software that operates on, or *processes,* incoming signals, usually on the audio signal before it's amplified for output to the speaker or headphones. More advanced DSP can act on the signals at radio frequencies. DSP can perform filtering functions to remove off-frequency signals, reduce or eliminate several different kinds of noise, or automatically detect and remove an interfering tone. Such abilities make DSP filtering much more flexible than the filtering accomplished by a crystal filter, but DSP performance is just becoming equivalent to that of a good fixed-width crystal filter. In general, the higher the number of bits specified for DSP and the higher the frequency at which the DSP functions are performed, the better the DSP processing performs. (Look in your radio's operating manual or product specification sheet for more information.)

Mobile and portable HF radios

Since the introduction of the Kenwood TS-50, a gold rush of radios has been designed for the portable and mobile operator. Each year sees more bands and better features crammed into these small radios. These radios are quite capable as base stations if space is limited at home or a dual home/portable station is desired. Many include coverage of VHF and UHF bands.

Be aware that because they are so small, these rigs have to make some compromises compared to the high-performance designs. The operator interface is, by necessity, menu-driven. This menu-driven interface makes some adjustments less convenient, although the most-used controls remain on the front panel. The smaller rigs don't include internal antenna tuners at the 100-watt output level as the larger rigs do.

Where can you fit a radio in your vehicle or boat? If you have an RV or a yacht, you may not have a problem, but in a compact car or an 18-foot runabout, the space issue is quite a challenge. Luckily, many radios designed for mobile use, such as the radio in Figure 12-1, have *detachable front panels,* sometimes called *control heads.* A detachable panel allows you to put the body of the radio under the dash or a seat or on a bulkhead. If you share the car or boat, get agreement on where to place the radio before drilling any holes.

Table 12-3 includes examples of several popular mobile/portable rigs and some of their features. Like their larger base station cousins, you have to consider many features and different sets of accessories. Because these radios are small and not all features have a dedicated front panel control, I recommend that you try one before you buy either at a dealers or with a friend who owns one.

Figure 12-1:
Most mobile
radios have
detachable
front panels,
as this radio
does.

Table 12-3		Mobile/Portable All-Mode HF Transceivers				
	Model	*VHF/UHF Coverage*	*Power Output*	*General Coverage Receive*	*Detachable Control Head*	*Receives FM Broadcast*
Alinco	DX-70	50 MHz	100 watts	Yes	Yes	No
Icom	IC-706 MKIIG	50/144/ 440 MHz	HF - 100w 50 MHz - 100w 144 MHz - 50w 440 MHz – 20w	Yes	Yes	Yes
	IC-703	50 MHz	10 watts	Yes	No	No

(continued)

Table 12-3 *(continued)*

	Model	VHF/UHF Coverage	Power Output	General Coverage Receive	Detachable Control Head	Receives FM Broadcast
Kenwood	TS-50	No	No	100 watts	Yes	No No
	TS-480	50 MHz	HF - 200w 50 MHz - 100w	Yes	Yes	No
Vertex Standard (Yaesu)	FT-100	50/144/440	HF - 100w 50 MHz - 100w 144 MHz - 50w 440 MHz - 20w	Yes	Yes	No
	FT-817	50/144/440	5 watts	Yes	No	Yes
	FT-857	50/144/440	HF - 100w 50 MHz - 100w 144 MHz - 50w 440 MHz - 20w	Yes	No	Yes
	FT-897	50/144/440	HF - 100w 50 MHz - 100w 144 MHz - 50w 440 MHz - 20w	Yes	No	No

Digital data on HF

More and more HF radios are providing a connector or two with a digital data interface to make connecting a personal computer and operating on the digital modes, such as PSK31 or RTTY, easier. A few even have a built-in data modem or a terminal node controller (TNC), which is a type of data modem used for packet radio (see Chapter 11). The key features to look for are accessory sockets on the radio carrying some of the following signals:

✔ **FSK (Frequency Shift Keying):** A digital signal at this connector pin causes the transmitter to output the two tones for frequency-shift keying, a method of transmitting using two frequencies, usually used for radioteletype (RTTY).

✔ **Data In/Out:** If a radio has an internal data modem, you can connect these digital data inputs and outputs to a computer. You may need an RS-232 (a type of serial communication) converter.

✔ **Line In/Out:** Audio inputs and outputs compatible with the signal levels of a computer's sound card, this input is used for digital data when a computer sound card is used as the data modem.

✔ **PTT:** The same as the push-to-talk feature on a microphone, this input allows a computer or other external equipment to key the transmitter.

✔ **Discriminator (sometimes labeled DISC):** This input is the unfiltered output of the FM demodulator. External equipment can use this signal both as a tuning indicator and to receive data.

To find out how to configure a radio to support digital data, look on the manufacturer's Web site or ask the dealer for the radio manual. Proper connections for RTTY (radioteletype) operation, packet radio, and other types of digital data should be included, and you can determine if your favorite mode is supported. If the manual doesn't provide an answer, contact the manufacturer to see if someone can tell you how to hook up the radio. Digital data Web sites may also have files on how to interface specific radios to your computer.

Making a decision on amplification

I recommend that you refrain from obtaining an amplifier for HF operations until you have some experience on the air. You need the extra savvy that experience provides to add an amplifier and then deal with the incumbent issues of power, feedlines, RF safety, and interference. The stronger signal you put out when using an amplifier also affects more hams if misadjusted or used inappropriately. Like learning to drive, perfecting the basic techniques is best accomplished before taking hot laps in a stock car.

Most HF radios output 100 watts or more, which is sufficient to do a lot of operating in any part of the hobby. When do you need a *full gallon* of 1500 watts output or even 500 to 800 watts? Many circumstances occur in which the extra punch of an amplified signal gets the job done. DXers use them to make contact over long paths on difficult bands. A traffic handler's amplifier gets switched on when a band is crowded or noisy so that the message gets through clearly. Digital operators use them to reduce the number of errors in received data. In emergencies, an amplifier may get the signal through to a station with a poor or damaged antenna.

HF amplifiers come in two varieties: vacuum tube and solid-state. Tubes are well suited to the high power levels involved. Solid-state amplifiers, on the other hand, tend to be more complex, but require no tuning or warm-up; just turn them on and go. Tube amplifiers are less expensive per watt of output power than solid-state amps, but they are larger and tubes are more fragile.

Don't attempt to use CB "footlocker"-type amplifiers. Not only are these amps illegal, but they often have serious design deficiencies that result in poor signal quality.

VHF and UHF radios

Many HF radios also include 50, 144, and 440 MHz operations. The Kenwood TS-2000 goes all the way to 1200 MHz! This power makes purchasing a second radio just for VHF/UHF operating less of a necessity for the casual operator. Many ham radio shacks have an all-band HF/VHF/UHF radio backed up with a VHF/UHF FM rig for using the local repeaters and packet radio.

VHF/UHF radios that operate in single-sideband (SSB), carrier wave (CW), and FM modes are known as *all-mode* or *multi-mode* to distinguish them from FM-only radios. If you get serious about operating on those bands and modes, then you may want to purchase a dedicated radio. VHF/UHF-only multi-mode rigs are less common than they used to be because many all-band radios now offer VHF/UHF coverage.

Many of the VHF/UHF all-mode radios have special features, such as full duplex operation and automatic compensation for transponder offsets, that make using the amateur satellites easier. (I introduce amateur satellite operation in Chapter 11.) Satellite operations require special considerations because of cross-band operation and the fact that they are moving, which results in a Doppler shift on the received signal.

An all-mode radio can also form the basis for operating on the amateur microwave bands. Commercial radios are not available for these bands (900 MHz; 2.3, 3.4, 5.6, 10, and 24 GHz; and up), so you can use a transverter instead. The *transverter* converts a received signal on the microwave bands to 28, 144, or 440 MHz bands where the radio treats it just like any other signal. Similarly, the transverter converts a low-power (100 milliwatts or so) output from the radio on back up to the higher band. Bringing the output signal up to 10 watts or more requires an external amplifier.

FM-only radios

FM on the VHF and UHF bands is used by nearly every ham regardless of their favorite operating style or mode. A newly minted Technician licensee can likely use an FM mobile or hand-held radio as his or her first radio. FM is available on the all-mode rigs, but because of the mode's popularity and utility, FM-only rigs are very popular. FM radios come in two basic styles: hand-held and mobile. You can use mobile rigs as base stations at home, too.

Hand-held radios

Hand-held radios come in single-band, dual-band, and multi-band models. With the multi-band radios covering 50–1296 MHz, why choose a lesser model? Expense, for one thing. The single-band models, particularly for 2-meters, cost less than half the price of a multi-band model. You do the lion's share of operating on the 2-meter (VHF 144–148 MHz) and 70-cm (UHF 430–440 MHz) bands, so the extra bands may not get much use.

You can expect the radio to include as standard features encoding and decoding of CTCSS sub-audible tones (tones used to restrict access to repeaters), variable repeater offsets, at least a dozen memory channels, and a DTMF keypad for entering control tones (similar to the tones available on a touch-tone telephone). A rechargeable battery and simple charger come with the radio. (I discuss repeater operation in Chapter 9.)

Extended-coverage receiving is a useful feature. I find being able to listen to broadcast FM and the low-VHF land mobile (public safety agencies, paging, and businesses use) bands around 2-meters is very useful. In addition, TV channels 2 through 13 can also receive between 54 and 216 MHz. That's a nice emergency feature, as well as entertaining.

What are your power output needs? The tiny credit-card size radios are convenient, but don't pack much of a punch. Unless you live in an area with excellent repeater coverage, insufficient power leaves you out of touch, perhaps when you really need it. Get a radio with at least 1 watt of output and pick up a spare battery, too.

Base and mobile radios

Mobile radios have a similar set of features as hand-held radios regarding memories, scanning, and controls. The more powerful transmitters used with an external antenna extend your range dramatically. Receivers in mobile radios often have better performance than those in hand-held radios — able to reject the strong signals from land mobile dispatch and paging transmitters on adjacent frequency bands. Information about how receivers perform in such conditions is available in product reviews in magazines such as *CQ* and *QST,* on the ARRL Web site (www.arrl.org), and on Web sites such as www.eham.net and www.qrz.com. Your own club members may have valuable experience to share, because they operate in the same places as you!

You can often use mobile radios for digital data operation on the VHF/UHF bands, particularly packet. When limited to 1200 baud data, as modem technology has advanced, hams have moved to use 9600 baud data. If you plan on using your mobile rig for digital data, make sure it is data-ready and rated for 9600 baud without modification.

Some rigs claim to be APRS or GPS compatible — what does that mean? First, check out the descriptions of the Automatic Position Reporting System (APRS) in Chapter 11. Location information from a GPS receiver is transmitted by packet to a network of APRS computers over VHF radio. A radio compatible with APRS has a serial port for the GPS receiver and an internal packet TNC that can send and receive APRS packets without any other external devices.

VHF/UHF amplifiers

Increasing the transmitted power from an all-mode, hand-held, or even small mobile radio is common. Amplifiers can turn a few watts of input into more than 100 watts of output. Solid-state commercial units are known as *bricks* because they are about the size of large bricks with heat sinking fins on the top. A small amp and external antenna can greatly improve the performance of a hand-held radio so that you don't need a separate mobile rig.

Amplifiers are either FM-only or SSB/FM models. Amplifiers just for FM use cause severe distortion of a single-sideband signal. An amplifier designed for SSB use is called a *linear amplifier* and SSB/FM models have a switch to change between the modes. You can amplify Morse code signals in either mode, with more gain available in the FM mode.

RF safety issues are much more pronounced above 30 MHz because the body absorbs energy more readily at those frequencies. An amplifier outputs enough power to pose a hazard, particularly if you use a beam antenna. Don't use an amplifier at 50 MHz or above if the antenna is close to people. Revisit your RF safety evaluation (see the *Ham Radio For Dummies* Web site) if you plan on adding a VHF/UHF amplifier to your mobile or home station.

Making a selection

Dozens of hand-held and mobile radios are for sale, so use a checklist of features to help you decide on a model. Note the capabilities you want as well as the ones that fall into the nice-to-have category. A checklist and comparison table is available on the *Ham Radio For Dummies* Web site to help you sort through the blizzard of features. The blank spaces are for you to add capabilities.

Choosing an Antenna

I can't say which is more important: the radio or the antenna. Making up for deficiencies in one by improving the other is difficult. A good antenna can make a weak radio sound better than the other way around. You need to give your antenna selection at least as much consideration as the radio.

This section touches on a number of types of useful and popular antennas. If you want to know more about antennas and want to try building a few yourself, you need more information. I can think of no better source for that information than *The ARRL Antenna Book,* now in its 20th edition. Not only a good ham resource, many professional antenna designers have a copy, too. Highly recommended! I include a list of useful antenna design books and Web sites in Appendix B.

HF antennas

At HF, antennas can be fairly large. An effective antenna is usually at least ¼-wavelength in some dimension. On 40-meters, for example, a ¼-wavelength vertical antenna is a metal tube or wire 33 feet high. At the higher HF frequencies, antenna sizes drop to 8–16 feet, but are still larger than even a big TV antenna. Your physical circumstances have a great effect on what antenna you can put up. Rest assured that a large variety of designs are available to get you on the air.

Wires, verticals, and beams are the three basic HF antennas used by hams all over the world. You can build all these antennas with common tools or purchase them from the many ham radio equipment vendors.

Wire antennas

The simplest wire antenna is a *dipole,* which is a piece of wire cut in the middle and attached to a feedline, as shown in Figure 12-2. The dipole gives much better performance than you may expect from such a simple antenna. To construct a dipole, use 10- to 18-gauge copper wire. It can be stranded or solid, bare or insulated. When completed, its length should be:

Length in feet = 468 / frequency of use in MHz

This formula accounts for a slight shortening effect that makes a ½-wavelength of wire slightly shorter than a ½-wavelength in air. For example, a dipole for 21.1 MHz is 468 / 21.1 = 22.2 feet long. Allow an extra 18 inches on each end for attaching to the end insulators and tuning and another foot (6 inches × 2) for attaching to the center insulator. The total length of wire you need is 22.2' + 18" + 18" + 12" = 26.2'.

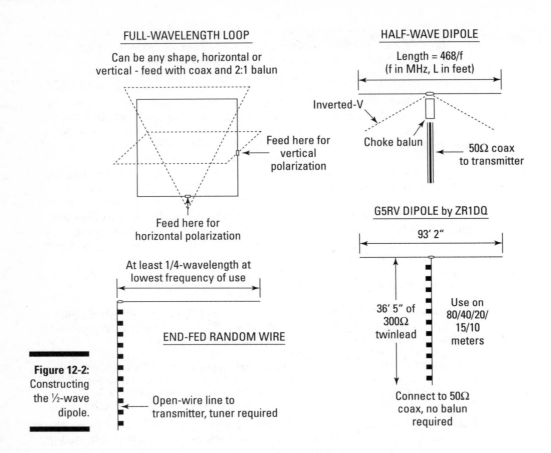

FULL-WAVELENGTH LOOP

Can be any shape, horizontal or
vertical - feed with coax and 2:1 balun

Feed here for
vertical
polarization

Feed here for
horizontal polarization

At least 1/4-wavelength at
lowest frequency of use

END-FED RANDOM WIRE

Open-wire line to
transmitter, tuner required

HALF-WAVE DIPOLE

Length = 468/f
(f in MHz, L in feet)

Inverted-V

Choke balun

50Ω coax
to transmitter

G5RV DIPOLE by ZR1DQ

93' 2"

36' 5" of
300Ω
twinlead

Use on
80/40/20/
15/10
meters

Connect to 50Ω
coax, no balun
required

Figure 12-2:
Constructing
the ½-wave
dipole.

To assemble a dipole, follow these steps:

1. **Cut the wire exactly in the middle and attach one piece to each end insulator, just twisting it back on itself for the initial check.**

2. **Attach the other end to the center insulator in the same way.**

3. **Attach the feedline at the center insulator and solder each connection.**

4. **Attach some ropes and hoist it up in the air.**

5. **Check the dipole.**

   Make some short, low power transmissions to measure the SWR (standing wave ratio) as explained in your radio's operating manual. The SWR should be somewhere less than 1.5 to 1 on the frequencies you wish to use.

6. **If the SWR is low enough at too high a frequency or is lowest at the high end of the band, loosen the connections at the end insulators and lengthen the antenna by a few inches on each end.**

   If the frequency of lower SWR is too low, shorten the antenna by the same amount.

7. **When you adjust the antenna length so that the SWR is satisfactory, make a secure wrap of the wire at the end insulators and trim the excess.**

You made a dipole! You can follow the same steps, except vary the lengths, for most simple wire antennas.

You can connect the dipole directly to the transmitter using coaxial cable and use the dipole on the band at which it is ¼-wavelength long or any odd number of ½ wavelengths. For example, the 66-foot long 7 MHz dipole works well on the 21 MHz band, too.

Other common and simple wire antenna designs include:

- **Inverted-V:** A dipole supported at its midpoint with the ends angling down at up to 45 degrees. This antenna only requires one support and gives good results in nearly all directions.

- **Full-wavelength loop:** Attach a feedline at the middle of the loop's bottom and erect the loop so that it is vertical. The feedline then works best broadside to the plane of the loop. These antennas are larger than the dipoles, but radiate a little more signal in their favored directions.

- **Multi-band dipoles:** Wires fed at the center with *open-wire* or *ladder-line* feedline and used with an antenna tuner to cover several bands. These are usually not ½-wavelength long on any band and so are called *doublets* to distinguish them from the ½-wavelength long dipoles.

- **Trap dipole:** Uses some appropriately placed components to isolate portions of the antenna at different frequencies so that the dipole acts like a simple ½-wavelength dipole on two or more bands.

- **The random-length wire:** Attach some open-wire feedline 15 to 35 feet from one end and extend the wire as far and high as you can. A couple of bends won't hurt. You have to convert the feedline to coaxial cable using a balun or antenna tuner. (I describe these on the *Ham Radio For Dummies* Web site.) The random wire's performance is hard to predict, but is an excellent back-up or temporary antenna.

For more information on these and many other antennas you can build, check out some of the antenna references listed in Appendix B. If you don't have the perfect backyard to construct the antenna of your dreams, don't be afraid to experiment. Get an antenna tuner (or use the one in your radio) and put up whatever you can. You can even bend wires or arrange them at strange angles. Antennas want to work!

Vertical antennas

Vertical antennas are nearly as popular as wire antennas. The ¼-wavelength and the ½-wavelength antennas are two common designs. Verticals don't require tall supports, keep a low visual profile, and are easy to move or carry. Verticals radiate fairly equally in all horizontal directions, so they are considered *omnidirectional* antennas.

The ¼-wavelength design is a lot like a ½-wavelength dipole cut in half and turned on end. The missing part of the dipole is supplied by an electrical mirror of sorts called a *ground screen* or *ground-plane*. A ground screen is made up of a dozen or more wires stretched out radially from the base of the antenna and laid on top of the ground. The feedline connects to the vertical tube (it can also be a wire) and to the radials, which are all connected together. The ¼-wavelength verticals are fairly easy to construct and, like dipoles, work on odd multiples of their lowest design frequency.

The ½-wavelength verticals are about twice as long as their ¼-wavelength counterparts, but do not require a ground screen. The lack of a ground screen means that you can mount them on masts or structures away from the ground and are ground-independent. The feedline is connected to the end of the vertical, but requires a special impedance matching circuit to work with low-impedance coaxial feedlines. Several commercial manufacturers offer ground-independent verticals and many hams with limited space or opportunities for traditional antennas make good use of them.

Both ¼-wavelength and ½-wavelength verticals can work on several bands through the use of the similar techniques used in wire antennas. Commercial multi-band verticals are available that you can use on up to nine of the HF bands.

Beam antennas

Any antenna that uses more than one *element* to focus or steer its listening and transmitting capabilities toward one direction is called a *beam,* which is short for *beamforming.* Beams can be as simple as two wires or as complicated as more than a dozen pieces of tubing.

The most popular type of beam antenna is called a *Yagi,* after Japanese scientists, Doctors Yagi and Uda, who invented the antenna back in the 1920s. The Yagi consists of two or more conducting elements (tubing or wires) parallel to each other and approximately 0.1 wavelength apart. The element that the feedline connects to is known as the *driven element.* Other elements called *reflectors* and *directors* are cut to specific lengths and spaced to reflect or direct the energy in a particular direction. Reflectors and directors, because they direct the energy without being directly connected to the feedline, are known as *parasitic elements,* and Yagi antennas are sometimes referred to as parasitic arrays. The most common ham Yagi today is a three-element design (a reflector, a driven element, and a director) that works on three popular ham bands (20, 15, and 10-meters) and so is called a *tri-bander.* Yagis are also made from wires at lower frequencies. Figure 12-3 shows a three-element Yagi beam on a 55-foot mast whose lowest operating frequency is 14 MHz.

Other beams are made from square or triangular loops. They work on the same principle as the Yagi, but with loops of wire instead of straight elements made from rod or tubing. Square-loop beams are called *quads* and the triangles are called *delta loops.*

You may encounter another type of beam that looks like an oversized TV antenna with many angled elements. These beams are *log-periodics,* or *logs,* and the large number of antenna elements enables the antenna to function over all bands in a certain range, just as TV antennas can receive many channels. Hams use logs to cover a range of bands (typically 20, 17, 15, 12, and 10-meters) with a single antenna.

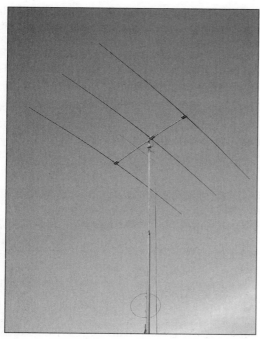

Figure 12-3:
My three-element beam antenna.

Whereas wire antennas have a fixed orientation and verticals radiate equally in all horizontal directions, a beam antenna made from aluminum tubing can be rotated. The ability to rotate the antenna allows the ham using a beam antenna to concentrate a signal or reject an interfering signal in a certain direction. You can place small HF beams on inexpensive masts or roof-top tripods, although they overload most structures designed for TV antennas. You also need a *rotator* that mounts on the fixed support and turns the beam. You can control the rotator from inside the shack and have a meter to indicate direction. (See the section, "Supporting Your Antenna," later in this chapter.)

Most hams start out with a wire or vertical antenna. After you operate for a while, the signals you hear on the air give you a good idea of what antennas are effective. After you have some on-the-air experience, you can make a decision as to whether you need a beam antenna.

VHF/UHF antennas

Most antennas used above 50 MHz at fixed stations are either verticals or beams. Verticals are used almost exclusively for FM operation, while beams are used for VHF/UHF DX-ing (contacting a distant station) on SSB and CW.

FM operating is done with *vertically polarized* antennas because of the mode's beginnings in the mobile radio industry. Antennas on vehicles are much simpler to mount vertically. In order to avoid cross-polarization losses (see the sidebar "Antenna polarization"), the base antennas were also verticals. This convention is nearly universal. Even if a beam antenna accesses a distant repeater, its elements are mounted vertically.

SSB and CW operators, on the other hand, use *horizontal polarization.* Many of the VHF and UHF propagation mechanisms respond best to horizontally polarized waves. If you have an all-mode radio and want to use it for both FM and SSB/CW, you need both vertically and horizontally polarized antennas available.

A popular and inexpensive vertical antenna is the simple *quarter-wave whip,* or *ground-plane,* antenna, which works the same as its larger HF cousin. Many hams build a 19-inch long 2-meter ground-plane as a first antenna project. You can extend the vertical to ⅝-wavelength versions, which offer a slightly stronger signal.

A ⅝-wavelength 2-meter vertical makes a dandy ¼-wavelength 6-meter vertical without any modifications!

TECHNICAL STUFF

Antenna polarization

Antennas radiate and receive *electromagnetic* fields composed of an electric and a magnetic field. Electrons in the conducting surfaces of an antenna move back and forth (creating a current) in the same direction as the oscillations of the electric field. This action causes currents to flow in the antenna, which either radiates a signal if the field is applied from a transmitter via a feedline or receives a signal if the field is caused by a distant transmitter.

The alignment of a radio wave's electric field with respect to the Earth's surface is called its *polarization.* Because the electric field and electron movement are parallel, an antenna's polarization is the same as the direction in which its conducting surfaces are arranged. Vertical antennas are *vertically polarized,* for example. Most Yagi antennas are *horizontally polarized.*

For the most efficient communications, you must orient the receive and transmit antennas so that the electromagnetic fields from each antenna are aligned. If the antennas aren't aligned, then the electric field from the transmitter does not make the receiving antenna's electrons move to make a current flow into the feedline. Such antennas are *cross-polarized* and the resulting loss of signal can be substantial. Polaroid sunglasses take advantage of the fact that most reflected light considered to be glare is horizontally polarized. By using vertically polarized plastic in the lenses, you can eliminate glare by cross-polarization.

Polarization is most important at VHF and UHF where signals usually arrive with their polarization largely intact. On the HF bands, the ionosphere scrambles the polarization so that relative polarization is much less important. Polarization does affect received signal strength, however, and deep signal fades caused by cross-polarization are common at HF.

Because of the reduced antenna size at higher frequencies, the Yagi antennas used by VHF/UHF DXers can have many more elements than at HF. Five to 7 element beams are common on 50 MHz and more than 20 elements are often seen at higher frequencies. These very high-gain antennas also have a narrow beamwidth, which is too narrow for most casual operating. If you choose to use a beam on these bands, 3 to 5 elements is a good choice on 6-meters and 5 to 12 elements at higher frequencies. These antennas are small enough to mount and turn using heavy-duty TV antenna hardware.

Mobile and portable antennas

For VHF, antennas for FM use are almost always vertical. The wavelength at these frequencies is such that a full-sized antenna is the norm. Mobile HF antennas, however, are generally reduced-size versions of verticals used at fixed stations. Horizontally polarized mobile HF antennas are rare due to their size and mechanical considerations. The challenge for mobile HFers is to get the most efficiency out of a short antenna.

You can mount mobile antennas as removable or permanent fixtures. The most easily removed antennas are the *mag-mount* models that use a magnetic base to hold themselves to a metal surface. Mag-mount antennas are available from HF through UHF. The drawback is that the installation isn't as clean as a permanently mounted antenna. Trunk-mount antennas for VHF and UHF are semi-permanent and look better. Drilling a hole in your car for a permanent mount looks best of all. All three options are fairly close in performance. Whichever method you choose, be sure you can remove the antenna from the mount to deter theft and for clearance, such as in a car wash. You can generally route antenna cables under trim, carpet, and seats.

For HF, many mobile stations use the "Hamstick"-type antennas attached to a mag-mount, as seen in Figure 12-4. These antennas are wire coiled on fiberglass tubes about 4 feet long, with a stainless-steel whip attached at the top. The antennas work on a single band and are sufficiently inexpensive that you can carry a whole set in the car. They require that you change the antenna when changing bands. Another design uses *resonators* attached to a permanent base to operate on different bands. You can attach several resonators at once so that several bands are used without changing the antenna. The resonators and fiberglass whip antennas use a standard ⅜-24 threaded mount.

An adjustable design has become popular in recent years with a moving top section that allows the antenna to tune over nearly any HF frequency. These antennas are known as *screwdriver* antennas because they use DC motors similar to those in battery-powered electric screwdrivers. In order of expense, the Hamstick-type antennas are the most inexpensive and the screwdriver antennas the most expensive. Performance varies dramatically depending on mounting and installation, so guaranteeing good results for any of the styles is difficult.

At VHF and UHF, portable antennas are very small, light, and easy to pack. At HF, however, the larger antennas are more difficult to deal with. You can try a lightweight wire antenna if you can find a way to support it above ground. Vertical antennas need a sturdy base and usually a ground system. Telescoping antennas may be an option, and you can use the mobile Hamstick-type whips with a few radial wires.

An antenna that does not present a low SWR (signal wave ratio, a measure of RF energy reflected back to the transmitter by the antenna) requires a tuner for the transmitter to output full power, adding weight and expense. The smaller coaxial feedlines, such as RG-58 types, also have higher losses. Try out the performance of your antenna and feedline before taking off on a major adventure. Avoid unpleasant surprises on the back roads!

Figure 12-4:
A "Hamstick"-type mobile antenna on one of my mobile adventures.

Feedline and connectors

Gee, how tough can picking a feedline be? It's just coax, right? Not really, and you may be surprised at how much feedline can affect your signal, both on transmit and receive. When I started out, I used 100 feet of RG-58 with a 66-foot dipole that I tuned on all bands. I didn't know that on the higher bands I was losing almost half of my transmitter output and received signals in the coax! The 50-foot piece I was using on my 2-meter antenna lost an even higher fraction of the signals.

While losing 50 percent or 3dB (decibels) is only about one-half of an S-unit (S-units are a measure of signal strength, roughly equivalent to 6dB each), I lost a few contacts when signals were weak or the band was noisy. To make up that loss with an amplifier costs around $300. Changing antennas to a beam with 3dB of gain costs at least that much, not counting the mast and rotator. That makes the extra $20 to $25 cost of using RG-213 cable look like a pretty good bargain!

Answering the decibel

Time for a refresher on decibels? Decibels, abbreviated dB, are used to measure a ratio of two quantities in terms of factors of 10. A change of a factor of 10 (from 10 to 100 or from 1 to 0.1) is a change of 10dB. Increases are positive and decreases are negative. You can use the following formula to calculate dB for changes in power and voltage:

    dB = 10 log (power ratio)
    dB = 20 log (voltage ratio)

Decibels add if ratios are multiplied together. For example, two doublings of power (x 2 x 2) is 3 dB + 3 dB = 6 dB. A gain of 20 can also be thought of as x 10 x 2 = 10 dB + 3 dB = 13 dB.

If you memorize these dB-ratio pairs, you can save yourself a lot of calculating, because a precise dB calculation isn't needed very often.

Power x 2 = 3dB	Power x 1/2 = -3dB
Power x 4 = 6dB	Power x 1/4 = -6dB
Power x 5 = 7dB	Power x 1/5 = -7dB
Power x 8 = 9dB	Power x 1/8 = -9dB
Power x 10 = 10dB	Power x 1/10 = -10dB
Power x 20 = 13dB	Power x 1/20 = -13dB
Power x 50 = 17dB	Power x 1/50 = -17dB
Power x 100 = 20dB	Power x 1/100 = -20dB

A change of one receiver S-unit represents approximately 6dB.

Table 12-4 compares several popular feedlines in terms of their relative cost (based on RG-58C/U) and the loss for a 100-foot section at 30 MHz and 150 MHz. The loss is shown in dB and in S-units on a typical receiver, assuming that one S-unit is equivalent to 6dB.

Table 12-4	Relative Cost and Loss of Popular Feedlines			
Type of Line and Characteristic Impedance	**Outside Diameter (in.)**	**Cost per Foot Relative to RG-58C/U**	**Loss of 100' at 30 MHz in dB and S-Units**	**Loss of 100' at 150 MHz in dB and S-Units**
RG-174A/U (50 ohms)	0.100	0.8	6.4 dB/1 S-unit	>12 dB/ >2 S-units
RG-58C/U (50 ohms)	0.195	1.0	2.6 / 0.5	6.7 / 1.1
RG-8X (50 ohms)	0.242	1.5	2.0 / 0.3	4.6 / 0.7
RG-213/U (50 ohms)	0.405	2.1	1.2 / 0.2	3.1 / 0.5
1" ladder line	½" or 1" width	0.6 to 1.3	0.1 / <0.1	0.4 / <0.1

The moral of the story is to use the feedline with the lowest loss you can afford. Open-wire feedline is a special case because you add an impedance transformer or tuner to present a 50-ohm load to the transmitter, incurring extra expense and adding some loss.

To save a lot of money on feedline, buy it in 500-foot spools from a distributor. If you can't afford to buy the entire spool, share the spool with a friend or two. Splitting the expense is an excellent club buy and can save more than 50 percent over buying cable 50 or 100 feet at a time. Do the same for coaxial connectors.

Beware of used cable unless the seller is completely trustworthy. Old cable is not always bad, but can be lossy if water has gotten in at the end or from cracks or splits in the cable jacket. If the cable is sharply bent for a long period, the center conductor can migrate through the insulation to develop a short or change the cable properties. (Migration is a particular problem with foam-insulation cables.)

Before buying used cable, examine the cable closely. The jacket should be smooth and shiny, with no obvious nicks, dents, scrapes, cracks, or deposits of adhesive or tar (from being on a roof or outside a building). Slit the jacket at each end for about 1 foot and inspect the braid, which should be shiny and show no signs of corrosion or discoloration whatsoever. Slip the braid back. The center insulator should be clean and clear (if solid) or white if foam or Teflon synthetic. If the cable has a connector on the end, checking the cable condition may be difficult. Unless the connector is newly installed, you should replace it, so ask if you can cut the connector off to check the cable. If you can't cut it off, you probably shouldn't take a chance on the cable.

The standard RF connectors used by hams are BNC, UHF, and N-type connectors. BNC connectors are used for low power (up to 100 watts) at frequencies through 440 MHz. UHF connectors are used up to 2-meters and can handle full legal power. N connectors are used up through 1200 MHz, can handle full legal power, and are waterproof when properly installed. (Photos or drawings of connectors are shown in vendor catalogs, such as The RF Connection catalog at www.therfc.com.)

Good-quality connectors are available at low prices, so don't scrimp on this important component. A cheap connector works loose, lets water seep in, physically breaks, or corrodes, eating up your valuable signals. By far, the most common connector you'll work with is the PL-259, the plug that goes on the end of coaxial cables. By buying in quantity, you can get silver-plated, Teflon-insulated connectors for just over $1 each, a bargain compared to the price if you buy them individually. Avoid the shiny nickel-plated connectors as they are difficult to solder.

To crimp or not to crimp? You can install a crimp-on connector quickly and get reliable service if you use it indoors in low-humidity, mild temperature environments. Crimping tools, or *crimpers,* are available for less than $50. If you have a lot of connections to make, a crimp-on connector may be a good choice. However, a properly soldered, silver-plated connector outperforms a crimped connector and can be used outside, too.

Supporting Your Antenna

Antennas come in all shapes and sizes — from the size of a finger to behemoths that weigh hundreds of pounds. The one thing that all antennas have in common is that they need to be clear of obstacles and kept away from ground level, which is where most obstacles are.

Before you start mounting your antenna, take a minute to review some elementary safety information for working with antennas and their supports. The short article at www.arrl.org/tis/info/pdf/0106091.pdf points out a few of the common pitfalls in raising masts and towers. For working with trees, aside from common sense about climbing, you may want to consult with an arborist.

Antennas and trees

Although Marconi used a kite for his early experiments, a handy tree is probably the oldest antenna support. A tree often holds up wire antennas, which tend to be horizontal or use horizontal support ropes. The larger rotatable antennas and masts are rarely installed on a tree, even on a tall, straight conifer, because of the mechanical complexity, likelihood of damage to the tree, and mechanical interference between the antenna and tree.

Nevertheless, for the right kind of antennas, a tree is sturdy, nice to look at, and free! The goal is to get a pulley or halyard into the tree at the maximum height. If you are a climber (or can find someone to climb for you), you can place the pulley by hand. Otherwise, you have to figure out some way of getting a line through the tree so you can haul up a pulley. You may be able to just throw the antenna support line over a branch. Bear in mind that when a line is pressing against the bark of a tree, the tree can rub and chafe against the line until the line breaks. (This catastrophe always happens at night, in a storm, or right before an important contact.) If the line stays intact, the tree tries to grow around the line, creating a wound, which makes raising or lowering the antenna impossible.

If you intend for the tree to support an antenna permanently, bringing in an arborist or a tree service professional to do the job right, using sturdy, adequately rated materials, is worth the expense. The Radioworks company has a good introduction to antennas in trees at www.radioworks.com.html.

Masts and tripods

A wooden or metal mast is an inexpensive way to support an antenna up to 30 feet above ground. If you are handy with tools, making a home-brew mast is a good project and numerous articles about their construction are in the ham magazines. Masts are good candidates to hold up wire HF antennas and VHF/UHF antennas, such as verticals and small beams. If you are just supporting a VHF or UHF vertical, you won't need a heavy support and can probably make a self-supporting mast that doesn't need guy wires. However, if you have high winds or the mast is subjected to a side load (such as for a wire antenna), then it needs to be guyed.

One commercially available option is the telescoping *push-up mast* designed to hold small TV antennas and often installed on rooftops. Push-ups come in sizes up to 40 feet with guying points attached. You can also construct masts by stacking short sections of metal TV antenna mast, but you have to add your own guying points. You can't climb either telescoping or sectional masts, so mounting the antenna and then erecting the whole assembly is up to you. You can also mount a section or two of stacking mast on a chimney in order to support a small vertical. Push-up and TV masts are available along with all the necessary mounting and guying materials from RadioShack, hardware, and home improvement stores.

One step beyond the mast is the roof-mount tripod. The lighter tripods are used for TV antennas and can hold small amateur antennas. Larger tripods can handle mid-sized HF beams. Tripods are good solutions in urban areas and in subdivisions that may not allow a ground-mounted tower. Tripods are available from several tower and antenna manufacturers.

Towers

By far the sturdiest antenna support is the tower. Towers are available as self-supporting (unguyed), multisection crank-up, tilt-over, and guyed structures with heights of 30 feet and up. They are capable of handling the largest antennas at the highest heights, but they are also substantial construction projects, usually requiring a permit to erect. Table 12-5 lists several manufacturers of towers.

The most common ham tower is a *welded lattice tower.* It is built from 10-foot sections of steel tubing and welded braces and you must guy or tie it to a supporting structure, such as a house, at heights of 30 feet or more. A modest concrete base of several cubic feet is required to provide a footing. Lattice towers for amateur use are between 12 and 24 inches on a side and can be used to construct towers well over 100 feet high. Lattice towers are sufficiently strong to hold several large HF beam antennas, if properly guyed.

Tilt-over towers are lattice towers hinged in the middle so that the top sections can pivot towards the ground using a winch. Because of mechanical considerations, tilt-overs are limited to less than 100 feet in height.

Crank-up towers are constructed from telescoping tubing or lattice sections. A hand-operated or motorized winch raises and lowers the tower with a cable and pulley arrangement. A fully nested crank-up is usually 20 to 25 feet high, reducing visual impact to the neighborhood, and you can climb it to work on the antennas. Crank-ups also usually have a tilting base that aids in transporting and erecting the tower. Crank-ups are unguyed and so depend on a massive concrete foundation of several cubic yards to keep their center-of-gravity below ground level to prevent tipping over when fully extended. You can install crank-ups in small areas where guying is not possible; they are available in heights of up to 90 feet.

Self-supporting towers are triangular in cross section, are constructed of truss-like sections like lattice towers, and rely on a large concrete base for center-of-gravity control. They are simpler and less-expensive than crank-ups. Available at up to 100 feet, they have similar carrying capacity to lattice towers. Mounting antennas along the length of a self-supporting tower is more difficult than for a lattice tower with vertical supporting legs.

Table 12-5	Tower Manufacturers			
Antenna	*Web Site*	*Lattice*	*Crank-Up*	*Self-Supporting*
U.S. Towers	Sold through Ham Radio Outlet		X	
Rohn Industries*	www.rohnnet.com/Index.htm	X		X
Trylon	www.trylon.com			X
Heights Tower	www.heightstowers.com	X	X	X
Universal Aluminum Towers	Sold through distributors			X

* Rohn was purchased by Radian, a Canadian company that plans on continuing to manufacture tower components to the Rohn specifications. As of publication, it's not clear if the Rohn name will be used. Rohn towers have been a staple of ham radio antennas for decades and the Rohn brand will still be around for a long time.

Be extremely careful when buying a used tower or mast. Unless it has been in storage, exposure to the elements can cause corrosion, weakening welds and supporting members. If disassembled improperly, the tower can be damaged in subtle ways difficult to detect in separate tower sections. A tower or mast that has fallen is often warped, cracked, or otherwise unsafe. Have an expert accompany you to evaluate the material before you buy.

You can construct self-supporting towers from unorthodox materials such as telephone poles, light standards, well casing, and so on. If it can hold up an antenna, rest assured that a ham has used it to do so at some time. The challenge is to transport and erect the mast, then climb it safely and create a sturdy antenna mounting structure at the top.

Regardless of what you decide to use to hold up your antennas, hams have a wealth of experience to share in forums such as the TowerTalk e-mail list. You can sign up for TowerTalk at `www.contesting.com`. The topics discussed range from mounting verticals on a rooftop, to which rope is best, to giant HF beams, and how to locate True North. The list's membership includes experts who can handle some of the most difficult questions.

Is it a rotor or rotator?

A *rotor* is a part of a helicopter and has nothing whatever to do with ham radio. A *rotator,* on the other hand, is a motorized gadget that sits on a tower or mast and points antennas in different directions. Rotators are rated in terms of wind load, which is measured in square feet of antenna surface. If you decide on a rotatable antenna, you need to figure out its wind load in order to determine the size rotator it requires. Wind load ratings for antennas are available from the antenna manufacturer. The most popular rotators are made by Hy-Gain (`www.hy-gain.com`) and Yaesu (`www.yaesu.com`).

You need to be sure that you can mount the rotator on your tower or mast; some structures may need an adapter. Antennas mount on a pipe mast that then sits in a mast clamp on top of the rotator. If you mount the rotator on a mast, as on the left in Figure 12-5, you must mount the antenna at the top of the rotator to minimize side loading. In a tower, the rotator is attached to a *rotator plate* (a shelf inside the tower for the rotator to sit on) and the mast extends through a *sleeve* or *thrust bearing* (a tube or collar that hold the mast centered above the rotator), as at the center and right of Figure 12-5. Because using a bearing in this way prevents any side loading of the rotator, you can mount antennas well above the tower top if the mast is sufficiently strong.

An indicator assembly called a *control box,* which you install in the shack, controls the rotator. The connection to the rotator is made with a multi-conductor cable. Install the feedlines to the antennas in such a way so that they can accommodate the rotation of the antennas and mast.

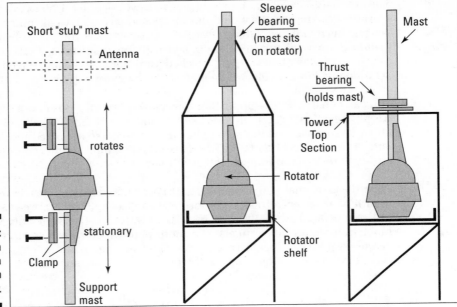

Figure 12-5: Mounting a rotator on a mast and on a tower.

Used rotators can be risky purchases. They are always installed in exposed locations and wear out in ways not visible externally. Even if the rotator turns properly on the bench, it may jam, stall, or slip when under a heavy load. Either buy a new rotator or get a used one from a trusted source for your first installation.

Radio accessories

You can buy or build hundreds of different gadgets to enhance whatever style or specialty you choose. Here's some information on the most common accessories that you need to get the most out of your station.

Mikes, keys, and keyers

Most radios come with a hand microphone, although if you buy a used radio, the hand mike may be long gone or somewhat worn. The manufacturer-supplied hand mikes are pretty good and are all you need to get started. After you operate for a while, you may decide to upgrade.

If you're a ragchewer, some microphones are designed for audio fidelity with a wide frequency response. Net operators and contesters like the hands-free convenience of a headset with an attached *boom mike* held in front of your mouth. Hand-held radios are more convenient to use with a combined *speaker-microphone* combination accessory that plugs into your radio and clips to a shirt pocket or collar. Your radio manufacturer may also offer a premium microphone as an option or accessory for your radio. Heil Sound (www.heilsound.com) manufacturers a wide range of top-quality microphones and headsets.

The frequency response of a microphone can make a big difference on the air. If you operate under crowded conditions, the audio from a microphone whose response emphasizes the mid-range and higher frequencies is more likely to cut through the noise. Some microphones have selectable frequency responses so that you can have a natural-sounding voice during a casual contact, then switch to the brighter response for some DX-ing. If you're not sure which is best, ask the folks you contact or do an over-the-air check with a friend who knows your voice.

Morse code enthusiasts have thousands of keys to choose from, spanning over a century of history (see Chapter 10 for more info). Beginners often start with a straight key and then graduate to an electronic keyer and a paddle. If you think you'll use CW a lot, I recommend going the keyer/paddle route right away. Many rigs now include a keyer as a standard option. You can plug the paddle into the radio and you're on your way! CW operators tend to find paddle choice very personal, so definitely try one out before you buy. A hamfest often has one or more key-bug-paddle collectors and you can try many different styles. The ham behind the table is likely to be full of good information, as well.

If you decide on an external keyer, you can choose from kits and finished models. Programmable memories are very handy for storing commonly sent information, such as your call sign or a CQ message. Sometimes I put my keyer in beacon mode to send a stored CQ message repeatedly to see if anyone is listening on a *dead* band. (If everybody listens and nobody transmits, the band sounds dead, but may be open to somewhere exciting.) A number of computer programs send code from the keyboard. Browse to www.ac6v.com/morseprograms.htm for an extensive listing of software.

What is a *voice keyer?* That sounds like an oxymoron! A voice keyer is a device that can store short voice messages and play them back into your radio as if you were speaking. Some are standalone units and some use a sound card. Voice keyers are handy for contesting, DX-ing, calling CQ, and so forth. Models also store both CW and voice messages, such as the MJF Contest Keyer (www.mfjenterprises.com) and the Super Combo Keyer (www.arraysolutions.com).

Antenna tuners

While your new radio may be equipped with an antenna tuner, a number of situations can occur in which you need an external unit. Internal tuners have a somewhat limited range that fits many antennas. Antennas being used far from their design or optimized frequency often present an impedance that the rig's tuner can't handle. If balanced feedlines are used, you may need a tuner that can handle the change from coaxial cable to open-wire feedlines.

Tuners are available in sizes from tiny, QRP-sized units to humongous, full-power boxes larger than many radios. Table 12-6 lists a few of the manufacturers offering an assortment of tuners. If you decide to purchase a tuner, choose one that's rated comfortably in excess of the maximum power you expect to use. I highly recommend getting one with the option to use balanced feedlines. The ability to switch between different feedlines and an SWR meter (which measures reflected RF power) are nice-to-have features.

They're called "antenna tuners" but they don't really "tune" an antenna at all! A more accurate name for these gadgets is an *impedance matcher*. The impedance matcher acts like an electrical gearbox to transform whatever impedance exists at the end of the feedline to the 50 ohms radios like to see. Your antenna doesn't even know it's there. This subject is discussed in detail in both *The ARRL Antenna Book* and *The ARRL Handbook*. In addition, the online article, "Do You Need an Antenna Tuner?" from *QST,* is online at `www.arrl.org/tis/info/tuner.html`.

Table 12-6	Antenna Tuner Manufacturers			
Manufacturer	*Web Site*	*Balanced Feedline*	*High-Power (>300 watts)*	*Automatic Tuning*
MFJ	www.mfjenterprises.com	Yes	Yes	Yes
Ameritron	www.ameritron.com	Yes	Yes	No
Vectronics	www.vectronics.com	Yes	Yes	No
LDG Electronics	www.ldgelectonics.com	External balun adapter	Yes	Yes
SGC	www.sgcworld.com	Yes	Yes	Yes
Alpha-Delta	www.alphadeltacom.com	Yes	No	Yes

Why is a rotator's wind load rated in square feet?

A rotator's job is to turn antennas and hold them in place against the pressure of the wind. The wind's pressure tries to turn the antenna against the rotator's braking mechanism. The rotator has to maintain control of the antenna when the brake is off and the antenna is turning against the wind. Both holding and turning the antenna require the rotator to exert torque on the mast that holds the antennas. As long as the rotator can exert more torque than the wind on the antenna, all is well. Torque caused by the wind on the antenna is difficult to specify exactly, but if a maximum wind speed is assumed, then torque is directly related to antenna area, which is easy to specify. Thus, rotators are rated in square feet of wind load. If multiple antennas are mounted on a single rotator, you must add their individual wind loads together. Rotators for amateur service are available with ratings starting at 3 square feet.

Along with the tuner, you need a *dummy load*. The dummy load is a large resistor that can dissipate the full power of your transmitter. The MFJ 260C can dissipate 300 watts, which is adequate for HF transceivers. High-power loads, such as the venerable Heathkit Cantenna or MFJ 250, immerse the resistor in a cooling oil. The dummy load helps minimize your transmitter causing interference during tune-up. HF dummy loads may not be suitable for use at VHF or UHF, so check the frequency coverage specification before you buy.

Computers in the Shack

As with just about every other activity, a computer can be involved. Ham radio is no different and has embraced computers more intimately than in most hobbies. Originally just used as a replacement for the paper logbook, the computer has evolved nearly to the point of becoming a second op: controlling radios, sending and receiving CW, and linking your shack to thousands of others through the Internet.

PC or Mac or . . . ?

Most ham shack computers are Windows-based machinery. The vast majority of software available for ham applications runs on the Win9*x*/2000/XP operating systems, with a healthy number of MS-DOS applications that run in MS-DOS windows on Win9*x*/2000.

Linux has an increasing number of adherents, particularly among the digital mode enthusiasts. Here are a few Web sites that focus on Linux software:

- **Hamsoft:** radio.linux.org.au
- **Terry Dawson VK2KTJ's listing:** www.radio.org/linux
- **AC6V.com:** www.ac6v.com/software.htm#LIN
- **Tucson Amateur Packet Radio:** www.tapr.org

The Macintosh computing community is making additional inroads to ham radio software and programs are available for all of the common ham radio uses available. The Ham-Mac mailing list is full of information for Mac fans (mailman.qth.net/mailman/listinfo/ham-mac). You can find Macintosh ham radio software at dogparksoftware.com and www.blackcatsystems.com. A useful Web site that is devoted to bringing together Macintosh computers and ham radio is www.machamradio.com.

Regardless of what platform and operating system you prefer, software tools and programs are available to help you enjoy any type of operating you like. While some software is supplied by commercial businesses, the amateur community has developed an amazing amount of shareware and freeware. Hams freely contribute their expertise in any number of ways, and developing software is a very popular activity.

Digital modes

Operation on most of the digital modes is rapidly converging on the sound card as the standard device to send and receive data. With a simple audio and transmit control interface, your computer and radio form a very powerful data terminal. Proprietary protocols, such as PACTOR II and Clover, require special hardware and software available from the protocol developer. I list specific software packages for the popular digital modes in Chapter 11. MFJ Enterprises (www.mfjenterprises.com) and West Mountain Radio (www.westmountainradio.com) both manufacture popular data interfaces.

If you choose to use an external multi-mode controller for the digital modes, such as the Timewave PK232 or DSP599 (www.timewave.com), Kantronics KAM (www.kantronics.com), or MFJ 1278B, you only need a terminal program, such as Hyperterm, which is built into Windows, or Symantec's Procomm-Plus (www.symantec.com). Numerous other terminal programs are optimized for ham radio digital data (which I list in Chapter 11).

Radio control

Radios are now provided with an RS-232 control interface through which you can monitor and control nearly every radio function. That flexibility has led to a number of radio control programs that put the front panel on a computer. Some radio manufacturers have a radio control package that you can purchase or download with the radio. A number of third-party programs integrate with logging software. If control interfaces are interesting to you, join a user's group for your radio on Usenet or via one of the ham radio portals (see Chapter 3).

Hardware considerations

Outside of computation-intensive applications, such as antenna modeling or high-performance data modems, you don't need to own the latest and greatest speed-demon computer. If you're thinking about upgrading a home computer, a computer that's a couple of years old does just fine in the ham shack. Furthermore, the flood of cheap, surplus computers available for a song means that you can dedicate a computer to its own specific task, such as running a packet node or monitoring an APRS Web site, so as not to tie up your main computer. Even a clunky old 486 DOS does just fine monitoring 1200-baud packet data, for example.

If you decide to purchase a new computer for the shack, be aware that the standard interface in ham radio for data and control is the RS-232 serial port. These ports are being phased out on new computers in favor of USB 2.0. (Serial ports are now referred to as *legacy* ports.) Integrating a USB-only PC into the ham shack means that you either have to purchase a serial port expansion card or use USB-to-RS-232 converters. The serial port expansion cards likely have fewer compatibility and driver issues, but the USB converters are easier to install.

Computer RF interference can also be an issue in the ham shack. The two main sources of interference are the monitor and data cables. If you have a noisy monitor — one that emits a lot of RF energy — the interference can be very tough to get rid of because shielding in monitors is almost non-existent. Ferrite cores are available from RadioShack (part numbers 273-104 and 273-105) that you can place on cables to choke off the interfering signals. Place the cores as close as possible to the connector on the device generating the interference. One core may be required at each end of the cable. Ground the metal case of any computing equipment, as well.

Buying New or Used Equipment

New equipment is always safest for a neophyte and it has that great new radio smell, too! If the equipment doesn't work, you have a service warranty or the customer service representatives can help you out. Sales personnel can help you with information about how to set up a radio or any accessories and may even have technical bulletins or application guides for popular activities. To find out where to buy new gear, get a copy of *QST, CQ,* or *WorldRadio.* The major dealers all run ads every month and some are virtual catalogs. For the smaller and local stores, check out the ARRL Technical Information Service's Web page at www.arrl.org/tis/tisfind.html.

While buying new equipment is safe, used gear is often an excellent bargain. Hams always love a bargain. You can find nearly any imaginable piece of gear with a little searching on a number of online swap sites, including eBay. (Look for these sites through the portals and reflectors listed in Chapter 3.) I like to buy and sell through the ham radio Web sites — my favorite is the Classified pages on the eHam.net portal (www.eham.net/classifieds). The ARRL and QRZ.com both have for-sale ads and quite a few other sites are springing up all the time. Enter **ham trade** or **ham swap** into an Internet search engine. As with shopping at hamfests, get help from an experienced friend before buying.

What about buying through mail order? Internet and toll-free telephone numbers make shopping for rock-bottom prices easy, but shopping by mail order has some drawbacks. The first is that you have to pay for shipping, which could add tens of dollars to your purchase. You may also have to pay for shipping to return equipment for service. Second, the local radio store is a valuable resource for you. RadioShack probably doesn't carry replacement parts for your WhizBang2000 nor does it probably have a new battery pack for that 10-Band PocketMaster. When you are stuck in the middle of an antenna or construction project, you don't want to have to stop and wait for UPS to deliver materials. My advice: Buy some things through the mail and some locally, balancing the need to save money against the need to support a local store.

Upgrading Your Station

Soon enough, usually about five minutes after your first QSO, you start thinking about upgrading your station to hear better or be stronger. Keep in mind the following tips when the urge to upgrade overcomes you. Remember the old adage, "You can't work 'em if you can't hear 'em!"

✔ The least expensive way to improve your transmit and receive capabilities is with better antennas. Dollar for dollar, you get the most improvement from an antenna upgrade. Raise them before making them larger.

✔ Only consider more power after you improve your antennas. Improve your hearing before extending the range at which people can hear you! An amplifier doesn't help you hear better at all.

✔ Buying additional receiving filters for your radio is a whole lot cheaper than a new radio.

✔ The easiest piece of equipment to upgrade in the station is the multimode processor between your ears. Before deciding that you need a new radio, be sure you know how to operate your old one to the best of your abilities. Improving your knowledge is often the cheapest and most effective improvement you can make.

By taking the improvement process one step at a time and by making sure that you improve your own capabilities and understanding, you can achieve your operating goals quicker and get much more enjoyment out of every ham radio dollar.

Chapter 13

Organizing Your Shack

. . . .

In This Chapter

▶ Devising your ham shack layout

▶ Staying safe with RF and electrical currents

▶ Grounding your equipment

. . . .

A well-organized shack provides a number of benefits for the occasional and serious ham enthusiast alike. You spend many hours in the shack, so why not make the effort to make your experience as enjoyable as possible? This chapter explains how to take care of the two most important inhabitants of the shack — the gear and you. The order of priority is up to you.

Designing Your Ham Shack

One thing that you can count on is that your first station layout will prove to be unsatisfactory. It's guaranteed! Don't bolt everything down right away. Plan to change the layout several times as you change your operating style and preferences.

You'll spend a lot of time in your shack, no matter where it's located, so making the operating comfortable and efficient is important. By thinking ahead, you can avoid some common pitfalls and save money, too!

Keeping a shack notebook

Before you unpack a single box or put up one shelf in your shack, you should start a shack notebook to record how you put your station together and to help you keep your station operating. The notebook can be a simple, spiral-bound book of notepaper or a three-ring binder. A bound book has the advantage that its pages can't get blown out of the binder due to an unexpected gust of wind.

Sketch out your designs before you begin building, make lists of equipment and accessories, and record the details as you go along. For example, if you hook up two pieces of gear with multi-conductor cable, write down the color of each wire and what it's attached to. You won't remember it later and the written list saves you tons of time and frustration.

After you have the station working, keep track of the gear you add and how you connect it. Record how your antennas work at different frequencies. Write down the wiring diagram for the little gadgets and adapters. Don't rely on memory! Taking a few minutes to record information saves the time tenfold later.

Building in ergonomics

Spending hours in front of a radio or workbench is common, so you need to have the very same concerns about ergonomics in your radio shack that you do at work. You want to avoid awkward positions, too-low or too-high furniture, and harsh lighting, to name just a few. By thinking about these things in advance, you can avoid any number of personal irritations entirely!

The focal point

Remember your main goals for the station. Building a station for HF ragchewing on phone, monitoring VHF FM repeaters, or using the digital modes requires a different approach for each. Whatever you plan on doing, you probably use one piece of equipment more than half the time. That equipment ought to be the focal point of how you arrange your station. The focal point can be the radio, a computer keyboard and monitor, or even a microphone or Morse code paddle. Paying attention to how you use that specific item pays dividends in operating comfort.

The computer

You may be building a radio station, but in many instances, you use your computer more than the radio! Certainly, the monitor is the largest piece of equipment at your operating position. Follow the guidelines for comfortable computer users. Position the desktop at the right height for extended periods of typing or, if possible, use a keyboard tray. Buy a high-quality monitor and place it at a height and distance for relaxed viewing. Figure 13-1 shows a few ways of integrating a monitor with radio gear.

Monitors mounted too far above the desk give you a sore neck. If you place the monitor too far away, your eyes hurt. Placed too far to the left or right, your back hurts. Now is the time to apply computer ergonomics and be sure that you don't build in aches and pains as the reward for the long hours you spend at the radio.

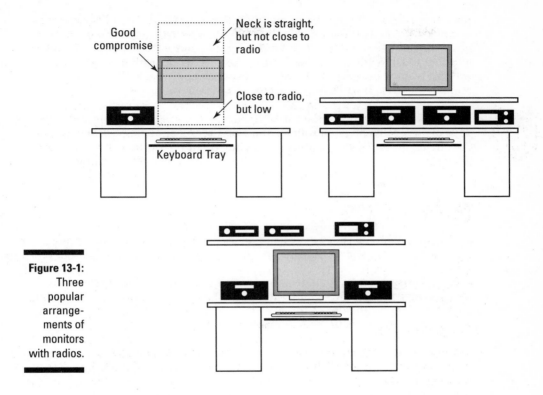

The radio

Radios (and operators) come in all different shapes and sizes, which makes giving hard and fast rules difficult. For example, HF operators tend to do a lot of tuning, so placement of the radio is very important. VHF-FM operators do less tuning, so the radio doesn't have to be as close to the operator. Placing your most-used radio off to one side of the keyboard or monitor probably is the most comfortable. If you are right-handed, have your main radio on your left and your mouse or trackball to the right. If left-handed, vice versa.

The larger radios usually have adjustable feet or a folding wire support called a *bail.* The bail folds flat against the underside of the rig for transport. Adjust the bail or feet to raise the front of the radio to a comfortable viewing angle. You should be able to see all of the control labels without having to move your head up or down.

The operating chair

A key piece of support equipment, so to speak, is the operator's chair. I am always astounded to visit state-of-the-art radio stations and find a cheap, wobbly, garage-sale office chair at the operating positions. Even though thousands of dollars were spent on electronics, the operator doesn't get the most out of the radios because of the chair.

A good shack chair is a roll-around office style chair with good lower-back support and plenty of padding in the seat. An adjustable model is best, preferably one that you can adjust with levers while sitting down. You may find that chairs with arms make sitting close to the operating desk difficult without leaning on your arms or stressing your lower back. If possible, remove the arms if you like sitting close to the operating table or desk.

The best shack chairs are the top-of-the-line secretarial models made for long days of easy swiveling and rolling. You won't regret spending a little extra on the chair. Your body is in contact with your chair longer than any other piece of equipment. You can find good bargains on used chairs at used office furniture stores, listed in the Yellow Pages.

The desk and shelves

The top surface of your operating desk is the second-most contacted piece of equipment. As is the case with chairs, many choices are available for desks suitable for a ham shack. Consider height and depth when looking at desks.

Before choosing a desk, you need to decide if the radio sits on the desk or on a shelf above it. Figure 13-2 illustrates the basic concerns. Do you like having your keyboard on the desk? You need to be comfortable sitting at your desk with your forearms resting comfortably on it. If you tune a radio a lot, such as for most HF stations, avoid arrangements that cause your arm to rest on the elbow or on the desk's edge for prolonged periods. Nothing is more painful! Make sure you have enough room for wrist support if using the keyboard is your main activity.

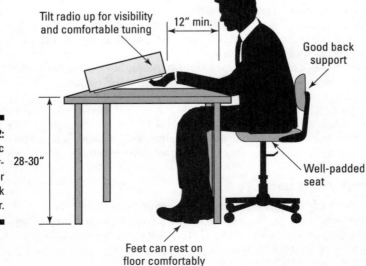

Tilt radio up for visibility and comfortable tuning

12" min.

Good back support

28-30"

Well-padded seat

Figure 13-2: The basic considerations for shack desk and chair.

Feet can rest on floor comfortably

The most common height for desks is between 28 and 30 inches from the floor. A depth of 30 inches is about the minimum if a typical transceiver that requires frequent tuning or adjustment is sitting directly on it. You need at least 12 inches between the front edge of the desk and the tuning knob. With your hand on the most frequently used radio control, your entire forearm needs to be on the desktop. If the radio is sitting on a shelf above the desk, be sure that it's close enough to you that tuning is comfortable and doesn't require a long reach. Be sure you can also see the controls clearly.

For small spaces, computer workstations may be a good solution. You probably have to add some shelves, but the main structure has all the right pieces and may be adjustable, to boot. One drawback is that they don't have much depth for arm support, if your operating style requires it.

Ham shack examples

Because every station location and use is going to be different, I can't provide hard and fast rules with which everyone must comply. The most helpful guidance I can provide is by way of example. Then you can decide what works for you. The goal of this section is to get you thinking about what works for you, not to suggest that you duplicate these stations exactly.

Paul Beringer NG7Z lives in a condominium where space is at a premium. He likes low-power HF contesting and DX-ing. His simple station, shown in Figure 13-3, resides in a self-assembly computer workstation with a fold-down front desk.

- ✔ The radio and accessories are stacked where Paul can easily see and operate them.

- ✔ The fold-down surface provides enough desk surface for wrist support when using a CW paddle.

- ✔ A slide-away drawer holds the computer keyboard and mouse below the desk at a comfortable height.

- ✔ The monitor is close enough to be easy on the eyes and within a comfortable viewing angle of the radio.

- ✔ Lighting is from behind the operating position to avoid glare.

Many hams like a soft light in the shack because it is less distracting and makes reading the indicators and displays on the radios and accessories easier.

Jack W1WEF has a larger area available for his station, adding a second transceiver and an amplifier to the mix. Jack spends a lot of time on the bands as you can tell from Figure 13-4. He often has two radios on at the same time.

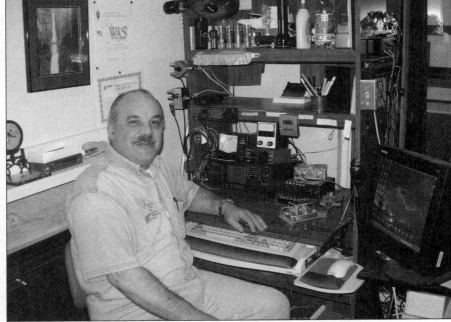

Figure 13-3:
Paul Beringer NG7Z has a lot of contacts in his log from this compact layout.

Figure 13-4:
Everything in this two-radio station is comfortably within reach of the operator.

✔ The computer monitor is right in the middle at a height that minimizes head and eye movement between the radios and software.

✔ Shelves above the monitor hold accessories and equipment that require few adjustments.

✔ Placing the amplifier on one side, antenna switches and rotators in the middle, and the power supply and speaker on the right minimize confusion and group similar controls together.

✔ Everything that requires frequent adjustment is within easy reach. No side-to-side movement is required.

Put paper labels on the front panel tuning controls of an amplifier to make changing frequencies easy without putting a powerful tune-up signal on the air. Easily changing frequencies minimizes stress on the amplifier and interference to other stations.

I use a computer cart as the foundation for my station, shown in Figure 13-5. I purchased a basic cart. Then I added plywood to extend the surface of the shelves to either side. A vertical piece of plywood on the left-hand side provides a mounting surface for switches to change antennas. I added another shelf at the cart's base for the computer and power supplies. Because I do almost all my work from the keyboard, it is at the most comfortable height for me with the main radio right behind it for easy reach.

Figure 13-5:
I have a rolling computer workstation with some added plywood shelves.

I placed the equipment that I don't adjust as much, including antenna rotators and VHF packet equipment, on a rolling cabinet to the left of the main station. The advantage of the rolling platforms is that I can move them easily to get access to the cables, something difficult to do with a fixed desk close to the wall.

Mobile and portable station examples

Getting a good ham station installed in a car has never been easier! In the past few years, manufacturers have taken the "all-band" radio design to new heights. Take the Icom IC-706mkIIG (or IC-706) for example. A radio only 6.6" x 2.3" x 7.9" and weighing only 5.5 pounds is capable of developing 100 watts output on all amateur bands between 1.8 and 50 MHz, 50 watts on 144 MHz, and 20 watts on 432 MHz. This radio and similar offerings from other manufacturers have changed mobile and portable operation forever.

The compact size of modern radios allows for very sleek installations in vehicles. You can even place radios with detachable front panels in the trunk or under a seat!

Dwayne Trego N8KDY took advantage of the space his unused ashtray was occupying in his car's dashboard. Removing the ashtray left plenty of room for his VHF/UHF mobile rig as you can see in Figure 13-6. Better yet, he even uses the ashtray cover to hide the rig when he's not in the car! Unless you're a ham, you probably wouldn't even notice the radio.

If you plan to operate from your RV, take a few pointers from the clean layout in Figure 13-7 by Pete Wilson K4CAV. When Pete stops at a campground, he sets up a station right at the driver's position. Pete not only has his HF transceiver mounted right on the dashboard, but it's sitting on top of a mobile amplifier for an extra-powerful signal. When in motion, all of this gear is safely stowed.

 Poorly secured radio gear can be lethal in an accident. Please don't go mobile without making sure your radios are securely mounted. Take care to keep cables and microphones away from air-bag deployment areas, as well. Anything in motion inside the vehicle is a hazard to you and your passengers.

Maybe you'd like something less tied to an automobile or RV. Well, how about a bicycle mobile station?

The bicycle mobile station, shown in Figure 13-8, is from Ben Schupack NW7DX, who constructed it during his senior year of high school. Ben uses a recumbent bike with an SGC-2020 HF QRP rig mounted on the handlebars. A whip antenna and batteries are mounted behind the rider.

Ready for Operating   Hidden by Ashtray Cover

Figure 13-6:
This clever installation replaces the car's ashtray with a VHF/UHF mobile rig.

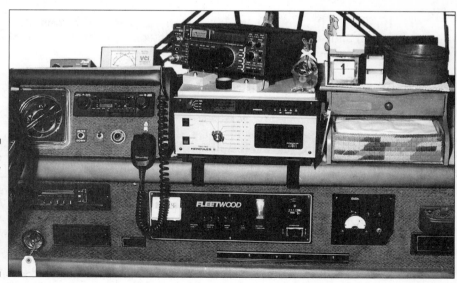

Figure 13-7:
Pete K4CAV operates right from the driver's seat when his RV is parked.

Figure 13-8:
Ben
Schupack
NW7DX
went for the
recumbent
look on
his two-
wheeler
setup.

Building in RF and Electrical Safety

Whatever type of station you choose to assemble, you must keep in mind basic safety principles. Extensive literature is available for hams (see the sidebar "Sources of RF and electrical safety information," later in this chapter).

Don't think that you can ignore safety in the ham shack. Sooner or later you get bit and equipment is damaged or someone gets hurt. Take a little time to review the safety fundamentals.

Basic safety

Safety is not particularly complicated. Mainly, being safe consists of consistently observing just a few simple rules. These tips can keep you out of more trouble than almost any others:

- ✔ **Know the fundamental wiring rules for AC power.** The National Electrical Code (NEC) contains the rules and tables that help you do a safe wiring job. The NEC, as well as numerous how-to and training references, is available in your local library or at home improvement centers. If you are unsure of your skills, hire an electrician.

- ✔ **Deal with DC power, especially in a car, carefully to avoid short-circuits and poor connections.** Either situation can cause expensive fires, and poor connections result in erratic operation of your radio. As with AC power, read the safety literature or hire a professional installer to do the job right.

✔ **Think of your own personal and family safety when constructing your station.** Don't leave any kind of electrical circuit exposed where someone can touch it accidentally. Use a *safety lockout* (devices that prevent a circuit breaker from being closed, energizing a circuit) on circuit breakers when you're working on wiring or equipment. Have fire extinguishers handy and in good working order. Show your family how to remove power safely from the ham shack.

Lightning

The power of lightning and its destructive potential is awesome. If you live in an area where lightning strikes are a possibility, take the necessary steps to protect your station and home. These steps can be as simple as disconnecting your antenna feedlines when not in use or you may decide to use professional-level bonding and grounding. Whatever you choose, do the job diligently and correctly.

RF exposure

Along with the possibilities of direct shocks due to contact with live electrical circuits, the signals your transmitter generates are also hazardous. The human body absorbs RF energy, turning it to heat. RF energy varies with frequency, being most hazardous in the VHF and UHF regions. A microwave oven operates at the high end of the UHF frequencies, for example.

Sources of RF and electrical safety information

Be responsible. Before you expose yourself to danger, check out these inexpensive safety references.

✔ The American Radio Relay League (ARRL) promotes safety for all manner of Amateur Radio activities. You can find excellent discussions of ham shack hazards and how to deal with them in the *ARRL Handbook* and the *ARRL Antenna Book*. A downloadable article is available in PDF format at www.arrl.org/tis/info/pdf/AntBk.pdf.

✔ For electrical safety issues relating to power circuits, the Electrical Safety Forum at

electrical-contractor.net/ESF/ Electrical_Safety_Forum.htm is a good source of information.

✔ Brush up on lightning and grounding issues in a series of Engineering Notes on the Polyphaser Web site (www.polyphaser. com). Click the Tech Info tab.

✔ The ARRL publication *RF Exposure and You,* the *ARRL Handbook,* and the *ARRL Antenna Book* all discuss RF exposure safety issues.

The ARRL provides a series of articles and guidelines covering all these safety issues at www.arrl.org/tis/tismenu.html.

Amateur signals are usually well below the threshold of any harmful effects, but can be harmful when antennas focus the signal in such a way that you're exposed for a long period of time. High-power VHF and UHF amplifiers can definitely be hazardous if you don't handle them with caution.

A comprehensive set of RF Safety guidelines is available. As you construct your own station, do a station evaluation to be sure you're not causing any hazards due to your transmissions. The *Ham Radio For Dummies* Web site has a downloadable form and procedure for you to use in evaluating your station's RF safety.

First aid

Just like engaging in any other hobby that involves the potential for injury, having some elementary skills in first aid is important. Have a first-aid kit in your home or shack and be sure other family members know where it is and how to use it. Training in first aid and CPR is always a good idea for you and your family, regardless of your hobby.

Grounding Power and RF

Providing good grounding — for both AC and DC power as well as for the radio signals — is very important for a trouble-free ham shack. Although proper power grounding is fairly straightforward for AC and DC power, grounding for the radio signals in a ham station is a different problem altogether. You have to deal with both.

Grounding for AC and DC power

Most people hear the word *ground* and think, "connected to the Earth." What the term really means, though, is the lowest common voltage for all equipment. For example, the Earth is at zero voltage for power systems. Grounding is really the process of making sure different pieces of equipment have the same voltage reference.

In AC and DC power wiring, grounding provides safety by connecting exposed conductors, such as equipment cases, either directly to the Earth or to a zero-voltage point, such as a building frame. Guiding any current away from you in the event of a short circuit between the power source and the exposed conductors provides safety. By keeping all equipment at the same low voltage, no current flows between pieces of equipment if they touch each other either directly or if you touch both pieces simultaneously.

Power safety grounding uses a dedicated conductor — the so-called "third wire." In the home, three-wire AC outlets connect the ground pin to the home's ground at the master circuit-breaker box. Because the frequency of AC power is very low, the length of the ground wire doesn't matter. It must only be heavy enough to handle any possible fault currents.

In DC systems, because of the generally low voltages involved (less than 30 volts), power safety is less concerned with preventing shock than with minimizing excessive current and poor connections. Both create a lot of heat and are significant fire hazards. You must pay careful attention to conductor size and keeping connections tight and clean. As with AC power grounds, the length of the conductor is not an issue; only its current-carrying capacity.

Grounding for RF signals

The techniques that work for AC and DC power safety often do not work well for the high frequencies that hams use. For RF, a wire doesn't have to be very long before it starts acting like an antenna or transmission line. For example, at 28 MHz, an 8-foot piece of wire is about ¼-wavelength long. It can have high voltage on one end and very little voltage on the other!

Because constructing a wiring system that has one common low-voltage point at RF for all the equipment is difficult, station builders use the more general definition of grounding, keeping all equipment at the same voltage. You can create an RF ground by using a single heavy wire or strap under or behind the equipment and tying each piece of gear to it with a short piece of strap or copper braid, as shown in Figure 13-9. Ham gear usually has a ground terminal just for this purpose.

Copper strap is sold at hardware and roofing stores as flashing. Avoid paying top dollar by finding a surplus metals dealer and poking around. A copper strip or sheet is often available in a wide variety of gauges and sizes. I was fortunate to find a heavy bar, predrilled with evenly spaced holes that made a dandy ground bus (see Figure 13-10). Use your imagination — all you need is material that is wide and is easy to make good electrical contacts with.

You don't have to buy expensive braid if you have some lengths of old coaxial cable around. As long as the coax is not waterlogged or corroded, the braid makes a fine ground strap. Lightly score the cable's outer plastic jacket and peel it off the cable. Push the ends of the braid sleeve together so that it loosens slightly. By pushing the braid off the center insulation, you can remove, flatten, and use it for grounding. The remaining wire is good high-voltage wire or you can use it for power supply applications.

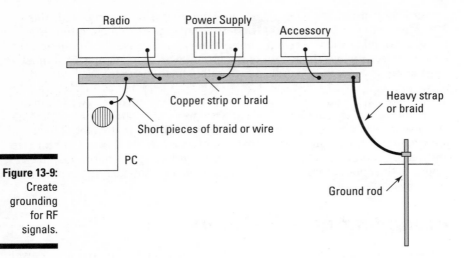

Figure 13-9:
Create grounding for RF signals.

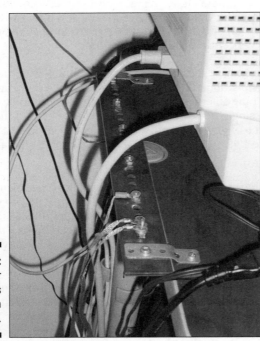

Figure 13-10:
A copper ground bus installed in my shack.

Chapter 14

Housekeeping (Logs and QSLs)

*B*efore you make any contacts with your new station, plan ahead for the housekeeping chores that go with a new shack. For the ham, keeping a log of station operations and sending QSLs for contacts are the paperwork that finishes the job. In this chapter, I show you what to put in your log and how to put together a QSL card.

Keeping a Log

Keeping a detailed log is no longer a requirement, but a lot of good reasons exist to keep track of what you do on the ham bands. The main log is a nice complement to the shack notebook (see Chapter 13), and you'll find it valuable in troubleshooting efforts and equipment evaluation.

Updating your little black radio book

Your station log can literally be a notebook or binder with handwritten entries for every contact. (Figure 14-1 shows a typical format for paper logbooks.) Be sure to record the basics:

▶ **Time:** Hams keep time in UTC (or World Time) for everything but local contacts and radiogram identification (see Chapter 10).

▶ **Frequency:** Just recording the band in either MHz or wavelength is sufficient (for example, 20-meters or 14 MHz).

▶ **Call sign:** Record each station you contact.

Record year somewhere on each sheet

Watts assumed   Use UTC!   RST optional   QSL sent & received

DATE	FREQ	MODE	POWER	TIME	STATION WORKED	REPORT SENT	REC'D	TIME OFF	QTH	COMMENTS NAME QSL VIA	QSL S	R
3 Nov	20 m	USB	100	1431	G3SXW	58	57	1437		Roger, funny chap	x	
				1438	OKIRI	57	58			Jiri	x	
				1442	DL6FBL	59	57			nr Munich, Ben	x	
				1515	V5IAS	59	59			Namibia, Ralph direct	x	
				1538	A6IAJ	59	59			UAE Ali K2UO	x	
4 Nov	146.82	FM	25	2000	N7UK	-	-	0224		ARES net		
				0225	N7FL	-	-			Debbie, needs battery		
8 Nov	40 m	LSB	100	0330	K8CC	57	55	Lots of QRN	MI	Dave		
	3.985	LSB	100	0405	W7EMD	59	57			WA Emergency Net		
9 Nov	10.124	PSK	100	0445	N7FSP	gud	OK			nice copy tonight		

Band or frequency OK

Can be used for any information about the QSO

Figure 14-1: A typical paper logsheet showing the basic information.

Those three pieces of information are enough to establish the who, when, and where of ham radio. Beyond the basics, you probably want to keep track of the mode you used, the signal report you gave and received, and any personal information about the other operator.

Don't limit yourself to just exchanging the information recorded in your log-book or logging program. Another person is on the other end of the QSO with lots of interesting things to say!

Keeping your log on a computer

If you're an active ham, I highly recommend keeping your log on a computer. The logging program makes looking up previous contacts with a station or operator easy. You can also use your logging program like a Web blog as a day-to-day radio diary to keep track of local weather, solar and ionospheric conditions, equipment performance, and behavior. The programs can also export data in standard file formats such as ADIF (Automated Data Interchange Format). For mobile or portable operation, portable logging programs, such as GOLog (`home.earthlink.net/%7Egolog/`), run on a Palm-OS PDA.

Selecting a QSL Card

QSL cards are the size of a standard postcard featuring an attractive graphic or photo with the station's call sign. Information about a contact is written on one side. The QSL is then mailed directly to the other station or through an intermediary as described below. One of my favorite activities, exchanging QSL cards allows you to claim credit for contacts in order to receive awards, but most hams just do it for fun.

You can find many varieties of QSLs, but three very basic rules can help make the exchange as quick and error-free as possible. Figure 14-2 illustrates the following:

✔ Have your call sign and QSO information on one side so that the receiver doesn't have to look for it.

✔ Print your call sign and all contact information in a clear and easy-to-read typeface.

✔ Beyond your mailing address, make sure the physical location of the station is shown, including county (for U.S. stations) and four-digit grid square.

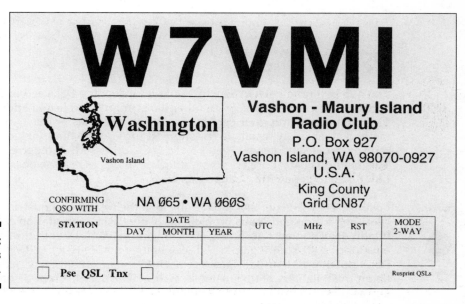

Figure 14-2:
My club's
QSL card.

You can find advertisements for QSL printers in the classified section of ham magazines or you can roll your own card with a laser or inkjet printer.

Follow these suggestions for sure-fire accuracy:

- **Double check the date and time of your contact.** Date and time are frequent sources of error. Start by making sure your own clock is set properly. Use UTC or World Time for every QSL except those for local contacts.

- **Use an unambiguous format for date.** Is 5/7/99 May 7th or July 5th? The date is crystal clear if you show the month with a Roman numeral, as in 7/V/99, or spell the month out as in 5 Jul 99.

- **Use heavy black or blue ink pen** that won't fade over time like colored inks. Never use pencil.

Sending and Receiving QSLs

After you fill out your QSLs, then what? Depending on where the cards are going, you have several options that trade postage expense against turn-around time. By following the appropriate rules for each method, you can get a high return rate for your cards.

QSL-ing direct

The quickest (and most expensive) option for sending QSLs is to send the cards *direct,* meaning directly to other hams at their published addresses. This method ensures your card gets to the recipient faster than any other method and usually results in the shortest turnaround time. Include the return postage and maybe even a *SASE* (self-addressed, stamped envelope). This method costs more for you, but makes things as easy as possible for the other ham to get a return card on its way.

Postal larceny is generally not a problem in the wealthier industrialized countries, but it is an enormous problem elsewhere. An active station can make hundreds of contacts per week, attracting unwelcome attention when many envelopes start showing up with those funny number-letter call sign things. Don't put any station call signs on the envelope if you have any question about the reliability of the postal service. Make your envelope as ordinary and as thin as possible.

Sending via managers

To avoid poor postal systems and cut postage expenses, many DX stations and nearly all DX-peditions use a *QSL manager*. The manager is typically in a country with good postal service and the return rate is nearly 100 percent. QSL-ing via a manager is just like QSL-ing direct. If you don't include return postage and an envelope to a manager for a DX station, you'll likely get your card back via the QSL bureau (see the next section), which takes a while. Managers can be located by a QSL manager Web site such as `www.qslinfo.de` or `www.eham.net`. The DX newsletters listed in Appendix B are also good sources of information.

If you send your QSL overseas, be sure to do the following:

✔ Use the correct Global Air Mail Letter Post rate from the U.S. Post Office Web site (`www.usps.com/common/category/postage.htm`).

✔ Ensure air mail service with an Air Mail sticker (free at the post office), an air mail envelope, or an Air Mail/Par Avion ink stamp on the envelope.

✔ Include return postage from sources such as William Plum DX Supplies (e-mail Bill at `plumdx@msn.com`) or the K3FN Air Mail Postage Service (`users.net1plus.com/ryoung/index.htm`).

An International Reply Coupon or *IRC* is redeemable for one unit of unregistered air mail postage. You can purchase IRCs at the post office or redeem one from overseas. If the clerk is not familiar with IRCs, section 372.1 of the USPS manual explains how to handle IRCs.

You'll hear "send 1 (or 2) greenstamps" for return postage. A *greenstamp* is a $1 bill. Take care that currency is not visible through the envelope. In some countries, non-domestic currency is illegal, so check before sending it through the mail.

Bureaus and QSL services

All that postage can mount up pretty quickly. A much cheaper (but slower) option exists: the QSL bureau system. You should use this method when the DX station says, "QSL via the bureau" or on CW, "QSL VIA BURO." The QSL bureau system operates as a sort of ham radio post office, which allows hams to exchange QSLs at a fraction of the cost of direct mail.

Bundle up all of your DX QSLs (you still have to send domestic cards directly) and send them to the Outgoing QSL Bureau. There the QSLs are sorted and sent in bulk to the Incoming QSL Bureaus around the world. The cards are then sorted and distributed to individual stations. The recipients then reciprocate and send reply cards back in the other direction. Go to `www.arrl.org/qsl/` for more details. To get your cards, you must keep postage and envelopes in stock at the Incoming QSL Bureau. Then, when you least expect it, a fat package of cards comes in the mail. What fun!

An intermediate route is the WF5E QSL Service (`www.qsl.net/wf5e/`), which forwards QSLs to foreign and U.S. managers at the cost of a dollar for a few cards. You send outbound cards directly to WF5E and your return cards come through the bureau system. Compared to direct mail, the WF5E QSL Service is still a good bargain.

QSL-ing electronically

Hey, this is the twenty-first century, isn't it? Why aren't hams sending QSLs electronically? Well, some hams are, although it's relatively new. Hams exchange cards on two different sites: eQSL and the ARRL's Logbook of the World (LOTW).

eQSL (`www.eqsl.net/qslcard/`) was the first electronic QSL system and is extremely easy to use. Its site has a tutorial slideshow that explains just how the eQSL works and how to use it.

The ARRL's LOTW (`www.arrl.org/lotw/`) is a little more complicated to use. You're required to provide proof of license and your identity in order to receive a *digital signature certificate.* (For information about how digital signatures work, look up references on Public Key Infrastructure technology.) The digital signature authenticates your contact information. LOTW does not generate or handle physical QSL cards; it only provides verification of QSOs for award purposes.

Chapter 15

Hands-On Radio

· ·

In This Chapter

▶ Acquiring tools and components

▶ Maintaining your station

▶ Troubleshooting

▶ Repairing your equipment

▶ Building equipment yourself

· ·

Ham radio is a lot more fun if you know how your radio works. You don't have to be an electrical engineer or a whiz-bang programmer, but to keep things running smoothly and deal with the inevitable hiccups, you need a variety of simple skills. As you tackle them, you'll find that you're having fewer problems, getting on the air more, and having better luck making contacts. Trying new modes or bands will also be much easier for you.

To help you get comfortable with the hands-on part of ham radio, this chapter provides some guidance in the three parts of keeping a ham radio station on the air: making sure your radio doesn't break, figuring out what is broken when it does, and fixing the broken part.

Before delving into the insides of your equipment, please take a minute and visit Chapter 13. Ham radio is a hobby, but electricity doesn't know that. I'd like to keep all of my readers for a long, long time, so follow one of ham radio's oldest rules, "Safety First!"

Acquiring Tools and Components

In order to take care of your radio station, you'll need to have some basic tools. It doesn't take a chest of exotic tools and racks of parts. In fact, you probably have most of the tools already. How many you need is really a question of how deeply you plan on delving into the electronics of the hobby.

You have the opportunity to do two levels of work: maintenance and repair or building.

Maintenance tools

Maintenance involves taking care of all your equipment, as well as being able to fabricate any necessary cables or fixtures to put it all together. Figure 15-1 shows a good set of maintenance tools. Having them on hand allows you to do almost any electronics maintenance task.

- **Wire cutters:** Use a heavy-duty pair to handle big wire and cable, and a very sharp pair of diagonal cutters, or *dikes,* with pointed ends to handle the small jobs.

- **Soldering iron and gun:** You need a model with adjustable temperature and interchangeable tips. Delicate connectors and printed-circuit boards need a low-temperature, fine-point tip. Heavier wiring jobs take more heat and a bigger tip. The soldering gun should have at least 100 watts of power for antenna and cable soldering. Don't try to use a soldering gun on small jobs.

- **Terminal crimpers:** Regular pliers on crimp terminals don't do the job; the connection pulls out or works loose and you spend hours chasing down the loose connections. Learn how to install crimp terminals and do it right the first time with the right tools.

- **Head-mounted magnifier:** Electronic components are getting smaller by the hour, so do your eyes a favor. Magnifiers are often available at craft stores. You can also find clamp-mounted swing-arm magnifier-light combinations.

- **Volt-ohmmeter (VOM):** If you can afford it, get one with diode and transistor checking, a continuity tester, and maybe a capacitor and inductance checker. Some models also include a frequency counter, which can come in handy.

You also need to have spare parts on hand. Start by having a spare for all your equipment's connectors. Look over each piece of gear and note what type of connector is required. When you're done, head down to the local electronics emporium and pick up one or two of each type. To make up coaxial cables, you need to have a few RF connectors on hand of the common types: UHF, BNC, and N.

You often need adapters when you don't have just the right cable or a new accessory has a different type of connector. Table 15-1 shows the most common adapter types. You don't have to get them all at once, but this list is good to have at a hamfest or to use when you need an extra part to make up a minimum order.

Headmount Magnifier

Volt-Ohmmeter   Soldering Iron and Tips

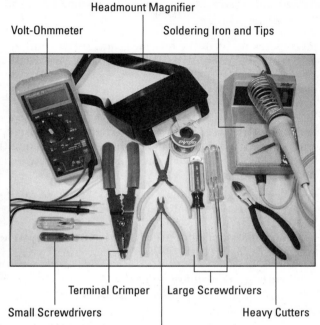

Figure 15-1:
A set of
tools
needed for
routine ham
shack main-
tenance.

Terminal Crimper   Large Screwdrivers

Small Screwdrivers   Heavy Cutters

Needlenose Pliers and Diagonal Cutters

Table 15-1	Common Shack Adapters
Adapter Use Common Types	
Audio	Mono to stereo phone plug (¼-inch and ⅛-inch), ¼-inch to ⅛-inch phone plug, right-angle phone plug, phone plug to RCA (phono) jack and vice versa, RCA double-female for splices
Data	9-pin to 25-pin D-type, DIN-to-D cables, null modem cables and adapters, 9-pin and 25-pin double male/female (gender benders)
RF	Double-female (barrel) adapters for all three types of connectors, BNC plug to UHF jack (SO-239) and vice versa, N plug to UHF jack and vice versa

What are the differences between plugs, jacks, sockets, males, and females? A *plug* is the connector that goes on the end of a cable. A *jack* is the connector that's mounted on equipment. A *male connector* is one in which the signal contacts are exposed pins (disregard the outer shroud or shell). A *female connector* has recessed sockets that accept male connector pins.

Along with adapters and spare parts, you should have on hand some common consumable parts:

✓ **Fuses:** Have spares for all of the fuse sizes and styles your equipment uses. Never replace a fuse with a higher value fuse.

✓ **Electrical tape:** Use high-quality tape such as Scotch 33+ for important jobs, such as outdoor connector sealing, and get the cheap stuff for temporary or throwaway jobs.

✓ **Fasteners:** Purchase a parts-cabinet assortment with #4 through #10 screws, nuts, and lockwashers. Some equipment may require the smaller metric-sized fasteners. You need ¼-inch and ⅜-inch hardware for antennas and masts.

✓ **Interference suppressors:** Having a couple of filters and ferrite cores around allows you to address interference problems quickly.

Cleaning equipment is an important part of maintenance and you need the following items:

✓ **Soft bristle brushes:** Old paint brushes (small ones) and toothbrushes are great cleaning tools. I also keep a round brush for getting inside tubes and holes.

✓ **Metal bristle brushes:** Light duty steel and brass brushes clean up oxide and corrosion. Brass brushes do not scratch metal connectors, but do damage plastic knobs or displays. Don't forget to clean a brush of corrosion or grease after the job.

✓ **Solvents and sprays:** I have a bottle or can of lighter fluid, isopropyl alcohol, contact cleaner, and compressed air. Lighter fluid cleans panels and cabinets gently and quickly, as well as removes old adhesive and tape. Always test a solvent on a hidden part of a plastic piece before applying a larger quantity.

Repairing and building tools

Figure 15-2 shows additional hand tools that you need when you begin doing your own repair work or building equipment. (I don't show the larger tools, such as drills and bench vises.)

Repairing and building go beyond maintenance in that you work with metal and plastic materials. You also need some additional specialty tools and instruments for making adjustments and measurements:

✓ **Wattmeter or SWR meter:** When troubleshooting a transmitter, you need an independent power measurement device. Many inexpensive models work fairly well (stay away from those in the CB shops; they often are not calibrated properly when used away from CB frequencies), but the Bird Model 43 is the gold standard in ham radio. Different elements, or *slugs,* are used at different power levels and frequencies. You can find both used meters and elements.

- ✔ **RF and audio generators and oscilloscope:** While you can do a lot with DC tests and a voltmeter, radio is mostly about AC signals. To work with AC signals, you need a means to generate and view them. If you're serious about getting started in electronics, go to the ARRL Web site (`www.arrl.org/tis/info/HTML/Hands-On-Radio/`) to find out how to purchase 'scopes and generators.

- ✔ **Nibbling tool and chassis punch:** Starting with a round hole, the nibbler is a hand-operated punch that bites out a small rectangle of sheet metal or plastic. Use a nibbler to make a large rectangular or irregular opening and then file the hole to shape. The chassis punch makes a clean hole in up to ⅛-inch aluminum or 20-gauge steel. These are not cheap, but if you plan on building regularly, the time savings and quality improvements are enormous.

- ✔ **T-handled reamer and countersink:** The reamer allows you to enlarge a small hole to a precise fit. A countersink quickly smoothes a drilled hole's edges and removes burrs.

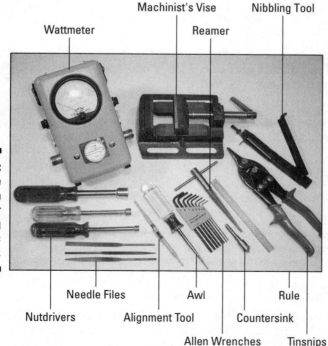

Machinist's Vise Nibbling Tool

Wattmeter   Reamer

Figure 15-2: Use these tools in building or repairing electronic equipment.

Needle Files Awl Rule

Nutdrivers Alignment Tool Countersink

Allen Wrenches Tinsnips

Drilling a hole in a panel or chassis that already has wiring or electronics mounted on or near it takes special measures. You must prevent metal chips from falling into the equipment and keep the drill from penetrating too far. To control chips, put a few layers of masking tape on the side of the panel where you are drilling with the outer layers kept loose to act as a safety net. To keep a drill bit from punching down into the wiring, use a small piece of hollow tubing that exposes just enough bit to penetrate all the way through.

Components for repairs and building

I find myself using the same components in the following list for most building and repair projects. Stock up on these and you'll always have what you need:

- **Resistors:** Various values of 5-percent metal- or carbon-film, ¼ and ½-watt fixed-value resistors; 100, 500, 1k, 5k, 10k, and 100k ohm variable resistors and controls

- **Capacitors:** 0.001, 0.01, and 0.1 µF ceramic; 1, 10, and 100 µF tantalum or electrolytic; 1000 and 10000 µF electrolytic; miscellaneous values between 220 and 10000 pF film or ceramic

- **Inductors:** 1, 10, and 100 mH chokes

- **Semiconductors:** 1N4148, 1N4001, 1N4007, and full-wave bridge rectifiers; 2N2222, 2N3904, and 2N3906 switching transistors; 2N7000 and IRF510 FETs; red and green LEDs

- **ICs:** 7805, 7812, 7912, 78L05, and 78L12 voltage regulators; LM741, LM358, and LM324 op-amps; LM555 timer; LM386 audio amplifier; MAX232 RS232 interface

Although having a completely stocked shop is nice, you'll find that building up the kinds of components you need takes time. Rather than give you a huge shopping list, here are some guidelines to follow:

- **When you buy or order components for a project, order extras.** The smaller components, such as resistors, capacitors, transistors, and diodes, are often cheaper if you buy in quantities of ten or more. After a few projects, you have a nice collection.

- **Hamfests are excellent sources of parts and component bargains.** Switches and other complex parts are particularly good deals. Parts drawers and cabinets often come with parts in them and you can use both!

- **Broken appliances and entertainment devices around the home are worth stripping before throwing out.** Power cords and transformers, headphone and speaker jacks, switches, and lots and lots of interesting hardware otherwise end up in the dump. Plus, seeing how they're made is interesting.

> ✔ **Build up a hardware junk box by tossing in any loose screws, nuts, spacers, springs, and so on.** Use an old paint tray or a flat, open tray to make rooting through the heap easy in search of a certain part. The junk box can be a real timesaver.

Maintaining Your Station

The best thing you can do for your station is to spend a little time doing regular maintenance. It works for cars, checkbooks, and relationships, so why not ham radio?

Be sure to keep a station notebook (see Chapter 13). The station notebook is a cheap and effective substitute for the frailties of human memory. Open the notebook whenever you add a piece of equipment, wire a gadget, note a problem, or fix a problem. Over time, the notebook helps you avoid or solve problems, but only if you keep it up to date.

You also need to set aside a little time on a regular basis to inspect, test, and check the individual components that make up the station. Along with the equipment, this includes the cables, power supplies, wires, ropes, masts, and everything else between the operator and the ionosphere. Check them when you plan to be off the air so that you don't have to do a panic fix when you want to be on the air. All the equipment and antennas in the world are of no use if they're not working.

You can make routine maintenance easy with a checklist. Start with the following list and customize it for your station:

> ✔ **Check all RF cables, connectors, switches, and grounds:** Make sure all connectors are tight because thermal cycling can work them loose. Rotate switches or cycle relays to keep contacts clean and turn up problems. Look for kinks in or damage to feedlines. Be sure that ground connections are snug.

> ✔ **Test transmitters and amplifiers for full power output on all bands:** Also double-check your antennas and RF cabling. Use full power output to check all bands for RF feedback or pickup on microphones, keying lines, or control signals.

> ✔ **Check received noise level (too high or too low) on all bands:** The noise level is a good indication of whether feedlines are in good shape, if preamps are working, or if you have a new noise source to worry about.

> ✔ **Check SWR on all antennas:** Be especially vigilant for changes in the frequency of minimum SWR, which can indicate connection problems or water getting into the antenna. Sudden changes in SWR (up or down)

mean tuning or feedline problems. SWR is discussed in Chapter 12 and in the technical content on the *Ham Radio For Dummies* Web site.

✔ **Inspect all antennas and outside feedlines:** Use a pair of binoculars or climb up and check. You're looking for loose connections; unraveling tape, ties, or twists; damage to cable jackets; and that sort of thing.

✔ **Inspect ropes and guys:** Get in the habit of checking for tightness and wear whenever you walk by. A branch rubbing on a rope can eventually cause a break. Knots can come loose.

✔ **Inspect masts, towers, and antenna mounts:** The time to find problems is well before the weather turns bad. Use a wrench to check tower and clamp bolts and nuts. Fight rust with a cold galvanizing paint. After the winter, check again for weather damage.

✔ **Vacuum and clean the operating table and equipment; clear away loose papers and magazines.** Sneak those coffee cups back to the kitchen and recycle the old soft drink cans. Make sure all fans and ventilation holes are clean and not blocked.

I realize that you may not want to haul the vacuum cleaner into the radio shack, but it may be the most valuable piece of maintenance gear you have. Heat is the mortal enemy of electronic components and leads to more failures than any other cause. The dust and crud that settles on radio equipment restricts air flow and acts as an insulator against heat dissipation. High voltage circuits, such as in an amplifier or computer monitor, attract dust like crazy. Vacuuming removes the dust, wire bits, paper scraps, and other junk so they can't cause expensive trouble.

As you complete your maintenance, note whether anything needs fixing or replacing and why, if you know. You will probably get some ideas about improvements or additions to the station, so note those, too.

Over time you will notice that some things regularly need work. In my station, because I have moveable desks, the ground connections need constant attention. After I noted regular trouble with a solid strap, I replaced it with a piece of braid and haven't had any more trouble. Regular maintenance uncovered the problem and I fixed it before I had an interference or feedback problem.

If you do routine maintenance three or four times a year, you can dramatically reduce the number of unpleasant surprises you receive.

Overall Troubleshooting Tips

No matter how well you do maintenance, something eventually breaks or fails. Finding the problem quickly is the hallmark of a master, but you can become a good troubleshooter by remembering a few simple rules:

✔ **Try not to jump to conclusions.** Work through the problem in an orderly fashion. Write your thoughts down to help focus.

✔ **Start at the big picture and work your way down to the equipment level.**

✔ **Avoid making assumptions.** Check out everything possible for yourself.

✔ **Read the equipment manual.** The manufacturer knows the equipment best.

✔ **Consult your station notebook.** Look for recent changes or prior instances of related behavior.

✔ **Write down any changes or adjustments while troubleshooting so you can reverse them later.** You may not remember everything that you did.

Troubleshooting Your Station

Your station is a collection of equipment (including antennas) connected together. To operate properly, each piece of equipment expects certain conditions to be met at each of its connectors and controls. You can trace many station problems to violations of those conditions, often without using any test equipment more sophisticated than a voltmeter. The problem usually is either the equipment or the connection.

Most station problems fall into two categories: RF and operational. RF problems are things such as high SWR, no signals, and reports of poor signal quality. Operational problems include not turning on (or off) properly, not keying (or keying inappropriately), or no communications between pieces of equipment.

Start by assigning the problem to one of these categories. (You may be wrong, but you have to start somewhere.)

RF problems

Some RF problems occur when RF is not going to where it's supposed to go. These problems are generally a bad or missing cable, connector, or switching device (a switch or relay) that needs to be replaced. Try fixing these problems with the following suggestions:

✔ **Replace cables and adapters if you have spares you know work.**

✔ **Note which combinations of switching devices and antennas seem to work and which don't.** See if the problem is common to a set or piece of equipment or specific cables.

✔ **Temporarily jumper around or bypass switches, relays, or filters.** If the jumper is not obvious, leave a note so you don't forget about it.

✔ **Check through antenna feedlines.** Take into account whether the antenna feedpoint has a DC connection across it, such as a tuning network or impedance matching transformer. Gamma-matched Yagi beams show an open circuit while beta-matched Yagis and quad loops have a few ohms of resistance across the feedpoint. (***Note:*** Recording the normal value of such resistances in the station notebook for comparison when troubleshooting is a good idea.)

✔ **Troubleshoot test equipment using the receiver's noise.** Disconnecting and reconnecting antenna cables causes changes in the noise level.

Other problems you may come across include hot microphones and equipment enclosures, or interference to computers or accessories. (You haven't fully lived until you get a little RF burn on your lip from a metal microphone case!) Usually you can fix these problems with a little grounding. Try these suggestions:

✔ **Double-check to ensure the equipment is grounded to the station RF ground bus.** The equipment may be grounded, but double-checking never hurts.

✔ **Check the shield connections on audio or control cables.** These cables are often fragile and might break when flexed or yanked. (You never yank cables, do you?)

✔ **Try different grounding connections or coil up an excessively long cable.**

✔ **Add ferrite RF suppression cores or beads to the cables to equipment you can't ground.** For more on this, skip ahead to the "Ferrites as RFI suppressors" sidebar.

On the higher HF bands, particularly 21, 24, and 28 MHz, connections begin to look like antennas as their lengths exceed ⅛ of a wavelength. For example, a 6-foot serial cable is about 3/16 wavelengths long on 28 MHz and has a sizeable RF voltage at the midpoint, even though both ends are grounded! If you have RF pickup problems on just one band, try attaching a ¼ wavelength *counterpoise* wire to move the RF hot spot away from the equipment in question. A quarter-wave conductor open-circuited at one end can look short-circuited at the other. Like balancing a tire with weights, attaching the counterpoise to the enclosure of the affected equipment may lower the RF voltage enough to reduce or eliminate the interference. Keep the wire insulated and away from people and equipment at the open end.

Operational problems

Operational problems fall into three categories: power, data, and control. After you determine which type of problem you have, you often come very close to the cause of the problem.

Power problems

Power problems can be obvious (no power), spectacular (high voltage power supply failure), or subtle (AC ripple, slightly low or high voltage, or poor connections). The key is to never take power for granted. Just because the power supply light is on doesn't mean the output is at the right voltage. I have wasted a lot of time due to not checking power and now I always check the power supply voltages first. Try these solutions to fix your power problems:

✔ **Check to see if the problem is caused by the equipment, not the power supply.** You can easily isolate obvious and spectacular failures, but don't just swap in another supply until you're sure that the problem is in fact the power supply. Connecting a power supply to a shorted cable or input can quickly destroy the supply's output circuits. If a circuit-breaker or fuse keeps opening, don't jumper it. Find the reason it's doing so.

✔ **Check for low output voltage.** Low voltage can cause all sorts of strange behavior by radios. The microprocessor may not function correctly, leading to bizarre displays, loss of external control, and incorrect response to controls. Low voltage can also result in low power output or poor RF stability (chirpy, drifting, or raspy signals).

✔ **Check the supply with both AC and DC meter ranges.** Hum on your signal can mean a failing power supply or battery. A DC voltmeter check may be just fine, but less than 100 mV of AC needs to be on the power supply output.

✔ **If you suspect a poor connection, measure voltage at the load (for example, the radio) and work your way back to the supply.** Poor connections in a cable or connector cause the voltage to drop under load. They can be difficult to isolate because they're only a problem with high current, such as when transmitting. Voltage may be fine when you receive. Excessive indicator light dimming is a sure indicator of poor connections or a failing power supply.

Working on AC line-powered and 50-volt or higher supplies can be dangerous. Follow safety rules and get help if you need it.

Data problems

Data problems are more and more common around modern radio shacks. Interfaces between computers, radios, and data controllers are usually made with RS-232 connections. If you installed new equipment and can't get it to play with your other equipment, three common culprits are to blame:

- ✔ **Baud rate:** An improper *baud rate* (or the data framing parameters of start bits, stop bits, and parity) renders links inoperative, even if the wiring is correct. Baud rate specifies how fast data is sent. The framing parameters specify the format for each byte of data. These may be set as part of a software interface or by switches in an accessory.

- ✔ **Protocol errors:** Protocol errors are generally a mismatch in equipment type or version. A PC using the Kenwood radio control protocol can't control a Yaesu or Ten-Tec radio, for example. Be sure all the equipment involved can actually use the same protocol or is specified for use with the exact models you have.

- ✔ **Improper wiring configuration:** Read the operating manuals for both pieces of equipment and be sure that you connect any required control signals properly. The cables may require jumpering pins at either or both ends. A common cable configuration, called a *null modem,* connects all RS-232 output signals to their companion input. You can make or buy null-modem cables or use a *straight-through* cable (pin 1 connected to pin 1, pin 2 to pin 2, and so on) and use a null-modem adapter at one end of the link.

RS-232 connections come in three common configurations. Three-wire configurations have Receive Data (RxD), Transmit Data (TxD), and ground — the software takes care of controlling the data flow. Five-wire configurations require RxD, TxD, ground, and Request to Send (RTS) and Clear to Send (CTS) control lines. The control lines implement *hardware handshaking* to control the data flow. If its control lines aren't configured properly, an output port won't send data. Seven-wire configurations add Data Terminal Ready (DTR) and Data Set Ready (DSR). You may also encounter Carrier Detect (CD) in packet data systems, indicating whether the channel is busy.

If you suddenly experience a failure in equipment that was communicating properly before, you may have a loose cable or the configuration of the software on one end of the link may have changed. Double-check the communications settings and swap cables.

You have a lot of options in wiring up an RS-232 link. Luckily, radio and accessory manufacturers generally don't get too fancy with the cabling, but you do need to read the manuals. If you'd like to know more about RS-232 interfaces, the tutorial at `www.data-com-experts.com/RS232_Data_Interface_tutorial.html` has a lot of useful information and connector wiring diagrams.

A *breakout box,* shown in Figure 15-3, is an invaluable tool for troubleshooting serial data connections. It allows you to monitor the status of all data and control lines. You can disconnect lines and jumper pins, as well. With a breakout box, determining if you have transmit and receive lines swapped or a control signal misrouted is easy. Less capable, but handy, are RS-232 status testers that show the state of each of several commonly used lines. Both are available from Jameco Electronics (`www.jameco.com`) and similar vendors.

New computers, as of early 2004, often have only USB or FireWire serial data ports. But the serial data interface standard in the radio shack is RS-232. You have two solutions: Install a multi-port RS-232 interface board in the computer or use USB-to-RS-232 adapters. Both require you to install drivers in your operating system. The interface boards tend to be the most compatible with radio gear. USB adapters convert the data twice (once between RS-232 and USB and then once between USB and the computer host), which leads to more opportunity for incompatibility. If you go the USB route, I suggest checking with the radio or accessory vendor for recommended adapters.

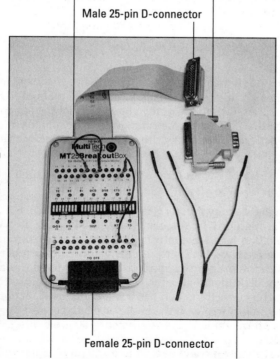

Jumper pins together   9-to-25 pin adapter

Male 25-pin D-connector

Figure 15-3: A breakout box allows you to view status and change configurations of serial data links.

Female 25-pin D-connector

One test pin for each connector pin   Wire jumpers

Control problems

Control problems are caused by either the infamous pilot error (in other words, you) or actual control input errors.

Pilot error is the easiest, but most embarrassing, to fix. With all the buttons and switches in the shack, I'm amazed I don't have more problems. Follow these steps to fix your error:

1. **Check that all of the operating controls are set properly.**

   Bumping or mistakenly moving a control is easy. Refer to the operator's manual for a list of settings for the various modes. Try doing a control-by-control setup and don't forget controls on the back panel or under an access panel.

   Speaking from hard personal experience, before you decide that a radio needs to go to the shop, check every control on the front panel, especially squelch (which can mute the audio unexpectedly), MOX (which turns the transmitter on all the time), and Receive Antenna (which makes the receiver sound dead if no receive antenna is attached). If you are really desperate, most radios have the capability of performing a *hard reset,* which restores all factory default settings but also wipes out the memory settings.

2. **Disconnect every cable from the radio one at a time, except for power and the antenna.**

   Start with the cable that contains signals related to the problem. If the behavior changes for any of the cables, dig into the manual to find out what that cable does. Could any of the signals in that cable cause the problem? Check the cable with an ohmmeter, especially for intermittent shorts or connections, by wiggling the connector while watching the meter.

3. **If the equipment is not responding to a control input, such as keying or PTT, then you need to *simulate* the control signal.**

   Most control signals are switch or contact closures between a connector pin and either ground or 12V. You can easily simulate a switch closure with . . . a switch! Replace the control cable with a spare, but unwired connector and use a test lead (wires with small clips on each end) to jumper the pin to the proper voltage. You may want to solder a small switch to the connector with short wires if the pins are close together. Manually make the connection and see if the equipment responds properly. If so, something is wrong in the cable or device generating the signal. If not, the problem is in the equipment you are testing.

At this point, you'll probably have isolated the problem to a specific piece of equipment, where your electronic skills can take over. You now have a decision to make. If you are experienced in electronics and have the necessary information about the equipment (schematic and operating manual), then by all means go ahead with your repairs. Otherwise, proceed with caution!

Troubleshooting Your Home and Neighborhood

If you have problems outside of your shack, they usually consist of the dreaded RF Interference (RFI), as in "I can hear you on my telephone!" or "My garage door is going up and down!" Lesser known, but just as irritating, is the man-bites-dog situation — your station receives interference from some other electric or electronic device. Solving these problems can lead you through some real Sherlock Holmes-ian detective work.

Start by browsing the ARRL RFI Information page located at www2.arrl.org/ tis/info/rfigen.html. For in-depth information, including diagrams and how-to instructions, read a copy of *The ARRL RFI Book,* which covers every common interference problem. Your club library may have a copy. Consult your club experts for assistance. Occasional interference problems are a fact of life in this modern era and you're not the only one to experience them. Use the experience and resources of others to help you out!

Dealing with interference to other equipment

Start by making your own home interference free. Unless you are a low-power VHF/UHF operator, you likely own at least one appliance that reacts to your transmissions by buzzing, humming, clicking, or doing its best duck imitation along with your speech. It's acting like a very unselective AM receiver and your strong signal is being converted to audio, just like the old crystal radio sets did. It's not the ham radio's fault — the appliance is failing to reject your signal — but it's still annoying.

Your goal is to keep your signal out of the appliance so that it doesn't receive the signal. Sounds simple, doesn't it? Start by removing all accessory cords and wires to see if the problem goes away. If it does, put the cords and wires back one at a time to see which one is acting as the antenna. Power cords and speaker leads are very good antennas and often conduct the RF into the appliance. Wind candidate cables onto a ferrite interference suppression core (RadioShack 273-104 and 273-105) close to the appliance and see if that cures the problem. You may have to core all the leads, although generally just one or two are sensitive. RadioShack also sells AC power cord filters (part number 15-1111) that may help.

Part 15 devices

Unlicensed devices that use RF signals to operate or communicate are subject to the FCC's Part 15 rules. These rules include cordless phones, wireless modems or headsets, garage door openers, and other such devices. Devices that may radiate RF signals unintentionally, such as computers and video games, are also subject to Part 15 rules. The rules make a tradeoff: The device owners don't need a license to operate the cordless phone, but are required not to interfere with and to accept any interference from a properly operating licensed service, such as Amateur Radio. This agreement generally works pretty well, except in the strong-transmitter/sensitive-receiver environment of a ham radio station. See the extensive discussion of Part 15 rules on the ARRL RFI Web site (www2.arrl.org/tis/info/rfigen.html) for more details.

If the device is battery-powered and doesn't have any leads, you probably can't fix the problem, I'm sorry to say. You have to either replace the device or get along with the interference. The manufacturer's Web site may have some interference cures, or you may find some guidance from the ham radio Web sites or club members. Try entering the model number of the appliance and **interference** into a search engine to see what turns up.

The following common devices are often victims of interference:

- ✔ **Cordless telephones:** Those that use 47 MHz frequencies are often devastatingly sensitive to strong out-of-band signals. Luckily, these phones are being replaced with models that use 900 MHz and 2.4 GHz radio links. These newer models are much less sensitive to your RF. If you come up against one of the 47 MHz units, just replace it with a newer one.

- ✔ **Touch lamps:** These accursed devices respond to nearly any strong signal on any frequency. You can try ferrite cores on the power cord, but results are definitely mixed. Internal modifications are described on the ARRL RFI Web site. Replacing the lamp may be the easiest option.

- ✔ **TVs and VCRs:** The usual point of entry for unwanted RF is through the VHF/UHF cable input. The sensitive tuners are easily overloaded. If the interference is from an HF signal, a high-pass filter often does the job. RadioShack 15-579 is an inexpensive filter that I use successfully. Cable installations may have loose connections that allow ham signals to leak in. The usual symptom is interference with cable channel 18 when transmitting on the 2-meter band. Have the cable company check and tighten the connections.

✔ **Alarm systems:** The many feet of wire strung around the house to the various sensors and switches make a dandy antenna. Unfortunately, the system controller sometimes confuses the RF they pick up for a sensor trip. System installers have factory-recommended interference suppression kits that take care of most problems.

By practicing on your own home electronics, you gain valuable experience in diagnosing and fixing interference problems. Also, if a neighbor has problems, you're prepared to deal with the issue. See the sidebar "Part 15 devices," in this chapter.

Dealing with interference to your equipment

You find two types of likely interference to you: electric and electronic. Electric noise is caused primarily by arcing in power lines or equipment, such as motors, heaters, and electric fences. Electronic noise is caused by leaking RF signals from consumer appliances and computers operating nearby or from nearby transmitters. Each has a distinctive *signature* or characteristic sound. The following list describes the signatures of common sources of electric noise:

✔ **Power line:** Steady or intermittent buzzing at 60 or 120 Hz; weather may affect interference.

Power line noise is caused by arcing or corona discharge. *Arcing* can occur around or even inside cracked or dirty insulators. It can also occur when two wires rub together, such as a neutral and ground wire. *Corona discharge* occurs at high-voltage points on sharp objects where the air molecules become ionized and electricity leaks into the atmosphere. The interference is a buzzing noise because the arc or discharge occurs at the peaks of the 60 Hz waveform, which occur at 120 Hz.

Do not attempt to fix problems with power lines. Always call your power company.

You can assist the power company by locating the faulty equipment. You can track down the noise source with a battery-powered AM radio or VHF/UHF hand-held radio with an AM mode (aircraft band works well). If you have a rotatable antenna at home, use it to pinpoint the direction of the noise. (The null off the side of a beam antenna is sharper than the peak of the pattern.) Walk or drive along the power lines in that direction to see if you can find a location where the noise peaks. I have found several power poles with bad hardware by driving around with the car's

AM radio tuned between stations. If you do find a suspect pole, write down any identifying numbers on the pole. Several numbers for the different companies that use the pole may be on it; write them all down. Contact your utility and ask to report interference. You can find a great deal more information about this process on the ARRL RFI Web page.

✔ **Industrial equipment:** Sounds like power line noise, but with a more regular pattern, such as the case with motors or heaters that operate on a cycle. Examples in the home include vacuum cleaners, furnace fans, and sewing machines.

✔ **Defective contacts:** Failing thermostats or switches carrying heavy loads emit highly erratic buzzing and rasping noise. These problems are significant fire hazards in the home and you need to fix them immediately.

✔ **Dimmers and speed controls:** Low-level noise like power lines that comes and goes as you use lights or motors.

✔ **Automotive ignition noise:** Buzzing that varies with engine speed, which is caused by arcing in the ignition system.

✔ **Electric fences:** Regular pop-pop-pop noises at about one second intervals. A defective charger can cause these problems, but the noise is usually due to broken or missing insulators or arcing from the fence wires to weeds, brush, or ground.

Ferrites as RFI suppressors

Ferrite is a magnetic ceramic material that is used as a core in RF inductors and transformers. It's formed into rods, *toroid cores* (circular and rectangular rings), and beads (small toroids made to slip over wires). Ferrite has good magnetic characteristics at RF and is made in different formulations, called *mixes,* that optimize it for different frequency ranges. For example, Amidon ferrites made of Type 73 material work best at HF and Type 43 ferrites work best at VHF.

Preventing RFI (RF interference) often means preventing unwanted RF current flow. Inductors have an increasing *reactance* (resistance to AC current) with frequency, so placing an inductor at the right spot can do the job. Forming the wire or cable into a coil does make an inductor, but getting enough inductance can mean making a sizeable coil, particularly below 10 MHz.

Winding the wire on a ferrite core or using a ferrite bead can create sufficient inductance in a very small volume.

Because of their small size, you can place ferrite cores and beads very close to the point at which an undesired signal is getting into or out of a piece of equipment. Beads are designed to slip over wires and cables. To create more inductance, use more beads. You can secure them on the cable with a plastic cable-tie, tape, or heat-shrink tubing. Cores are wound with several turns of the wire or cable. This technique works particularly well with telephone and power cords. *Split cores* come with a plastic cover that holds the core together, which makes placing the core on a cable or winding turns easy if the cable already has a large connector installed.

Finding an in-home source of electric noise depends on whether the device is in your home or a neighbor's. Tracking down in-home sources can be as simple as recognizing the pattern when the noise is present and correlating it to an appliance. You can also turn off your home's circuit breakers one at a time to find the circuit powering the device. Then check each device on that circuit.

If the noise is coming from outside your home, you have to identify the direction and then start walking or driving with a portable receiver. Review the ARRL RFI Web site or reference texts for information about how to proceed when the interfering device is on someone else's property.

What about electronic noise? No problem! The following list describes the signatures of common sources of electronic noise:

- **Computers or microprocessor-controlled games, entertainment systems, and appliances:** These produce steady or warbling tones on a single frequency, strongest on HF, but you can also hear them through VHF and UHF.

- **Cable and power-line modems:** You hear steady or warbling tones or hissing/rasping on the HF bands.

- **Cable TV leakage:** Heard at VHF and UHF, buzzing (video signal) or audio FM signal with program content. That TV may also experience interference from you on that cable channel.

- **Broadcast interference:** A steady AM signal with programming or bursts of speech or data interference is caused by overload of your receiver or by spurious signals generated from strong signals mixing together, and occasionally a spurious output from the broadcast transmitter itself.

Each type of electronic interference has its own set of techniques for finding the source and stopping the unwanted transmission. You are most likely to receive interference from devices in your own home or close by, because the signals are weak. If you are sure that the source is not on your property, you need a portable receiver that can hear the interfering signal.

The ARRL RFI Web site has some helpful hints on each type of interference as well as guidance on how to diplomatically approach the problem (because it's not your device). The ARRL RFI Web site's Overview page contains excellent material on helping you deal with and manage interference complaints (both by you and from others). ARRL members have the resources of the League's Technical Coordinators and technical information services, as well. You can completely eliminate or reduce most types of interference to insignificant levels with careful investigative work and application of the proper interference suppression techniques. The important thing is to keep frustration in check and work the problem through.

Building Equipment from a Kit

Building your own gear — even just a simple speaker switch — is a great ham tradition. By putting equipment together yourself, you become familiar with the operation, repair, and maintenance of your existing equipment.

If you're just getting started in electronics, I recommend that you start your building adventures with kits. When I got started, you could find Heathkit in every ham shack. Today, kits are available from many sources, such as Ramsey, Ten-Tec, Vectronics, RadioShack, and others. You can find numerous kit vendors on the ARRL Technical Information Service Web page (www.arrl. org/tis/) by clicking the TISfind link and entering **Kit** in the Search box.

Choose the simpler kits until you are confident of your technique. Kits are a great budget-saving way to add test instruments to your workbench and various gadgets to your radio station. Not only that, but you don't have to do the metal work and the finished result looks great, too! After you build a few kits, you'll be ready to move on up to building a complete radio. Although the Elecraft K2 (www.elecraft.com) is the top-of-the-line radio kit available today, numerous smaller QRP radio kits are available from other vendors.

You can build most kits using just the maintenance tool kit. Concentrate on advancing your soldering skills. Strive to make the completed kit look like a master built it and take pride in the quality of your work. Read the manual and use the schematic to understand how the kit works. Observe how the kit is put together mechanically, particularly the front panel displays and controls.

Building Equipment from Scratch

Building something by starting with a blank piece of paper or a magazine article and then putting it to use in your own station is a real thrill. Building from scratch is not too different from building a kit . . . except that you have to make your own kit. Your first project should be a copy of a circuit in a magazine or handbook — one that is known to work and gives directions on how to assemble and test it. If a blank printed circuit board (or *PCB*) is available, I recommend ordering one.

Imagine that you have to make a kit for someone else based on the instructions, schematic, and list of components. Photocopy the article and highlight all of the instructions. If an assembly drawing is included, enlarge it for guidance. Make extra copies so you can mark them up as you go. Read the article carefully to identify any critical steps. When you get your components together, sort them by type and value and place them in jars or an old muffin pan. Keep a notebook handy so that you can take notes for later use. As you build and test the unit and finally put it to use, everything is completely documented.

If you choose to design a circuit from scratch, I salute you! Documenting your work in a notebook is even more important to a project that starts with design. Take care to make your schematics complete and well-labeled. Record whatever calculations you must make so that if you have to revisit some part of the design later, you have a record of how you arrived at the original values. With digital cameras in abundance, take a few photos along the way at important milestones of construction. After you finish, record any tests that you make to verify that the equipment works.

Don't let failure get you down! The first cut at designs hardly ever work out exactly right and sometimes you even wind up letting all of the smoke out of a component or two! If a design doesn't work, figure out why and then move on to the next version. Don't be afraid to ask for help or try a different angle. Ham radio isn't a job, so keep things fun . . . after all, it's amateur radio!

Part V
The Part of Tens

The 5th Wave   By Rich Tennant

ADVANCED COMMUNICATIONS IN THE LUPINE COMMUNITY

"CQ...CQ...CQ...AOOOOOOOW!"

In this part . . .

In this part, I dispense more ham wisdom, ten tips at a time. I start with ten things I wish I'd known when I started in ham radio. Then I part the curtains of mystery and blab out ten secrets of the masters. Consider each one a coupon good for one helping of pure ham radio enjoyment!

Chapter 16

Ten Secrets for Beginners

In this chapter, I present ten fundamental truths that can help even the rankest beginner keep the wheels turning during those first forays into ham radio. Keep these tips in mind and you'll be on your way to veteran status in no time.

Listening, Listening, Listening

Listening is the most powerful and important way to learn. Listen to the successful stations to learn their techniques. Listening to on-the-air contacts is called *reading the mail*. All ham communications are open and public — they can't be encrypted or obscured. Just turn on the radio and get a real-time seminar in any facet of ham radio communication techniques you care to try.

Buddying Up

Find a friend who is learning the ropes like you are. Is there someone from your licensing class or club also getting started? Meet on the air and get used to using your equipment together. The best part is sharing in each other's successes!

Knowing Your Equipment

Hams joke about never reading the owner's manual, but don't believe it. Hams need to know their equipment. If a demo or tutorial is available to you, go through it. Practice adjusting the main controls or settings to observe the effects. Acquire at least a passing familiarity with even the most obscure controls. Keep the manual handy for quick reference, too.

Following the Manufacturer's Recommendations

The manufacturers want you to get the best performance and satisfaction out of their equipment, don't they? That's why they have recommended settings and procedures. Follow them until you are comfortable enough to optimize performance on your own.

Trying Different Things

Don't feel like you have to stay with one mode or band or magazine or radio. Changing your mind and striking out in a different direction is okay. As you become more comfortable with ham radio, feel free to dabble in anything that catches your fancy. Sooner or later, you'll discover something that makes you want to dive in deeply.

Nobody Knows Everything

Surround yourself with handbooks and how-to articles, magazines, Web sites, manuals, and catalogs. Use any reference available. If you're confused or not getting the results you expect, ask someone at a club meeting, on the air, or in an Internet forum for help. The oldest tradition in ham radio is hams giving other hams a hand. We're amateurs! We like to do it! Someone helped us, and we'll help you.

Practicing Courtesy

Behind every receiver is a person just like you. Polite terms like "Please," "Thanks," "Excuse me," and "Sorry" work just as well on the air as they do in person. Listen before transmitting and be flexible. If you encounter a rude operator, just go somewhere else or find something else to do — don't let tempers escalate on the air.

Joining In

By definition, ham radio isn't a solitary activity. Ham radio is a lot more fun if you have some regular acquaintances. Being welcomed on the air into a round-table QSO or a local weather net is great. Ham radio welcomes kings and paupers equally and we're all on a first-name basis. The ham bands are your home.

Getting Right Back in the Saddle

So what if you called CQ and nobody responded? So what if you put up a new antenna and it didn't work? Get right back in the saddle and try again. I don't know of any ham who has instant success right off the bat, so don't get discouraged and give up. You worked too hard to get that license!

Relax, It's a Hobby!

I know the scary feeling of thinking every ham is listening whenever you get on the air. Hey, relax and don't worry about a mistake putting you on a hobby-wide blacklist. If you try something new and it doesn't work out, that's okay. Everybody fumbles now and then. Keep ham radio fun for yourself and do things you enjoy.

Chapter 17

Ten Secrets of the Masters

Surely, the grizzled veterans have stores of secret knowledge that take years and years to learn and make them the masters of all they survey? Certainly they have their experiences and expertise, but down deep, you can find they rely on simple principles that work in many different situations. You can use them, too!

Listening, Listening, Listening

The masters get more out of listening than anyone because they have learned how to listen. Every minute you spend listening is a minute learning and a minute closer to being a master.

Learning What's Under the Hood

Operating a radio and building an efficient and effective station are so much easier if you know how the equipment works. Even if you're not terribly tech-savvy, take the time to learn the basics of electronics and how your equipment functions. A master understands the effects of controls and adjustments.

Reading History

An appreciation of the rich history of ham radio helps you understand how the hobby has shaped itself. Ham radio is full of conventions and methods developed over time, many of which can seem at first confusing and obscure. Master these conventions by understanding how they came about.

Having a Sharp Axe

When asked what he would do if given eight hours to cut down a tree, Abe Lincoln replied that he would spend the first six hours sharpening his axe. Masters keep their equipment and skills sharp. When they're needed on the air, they're ready.

Practicing Makes Perfect

Even a sharp axe gets dull over time. A master is on the air regularly, keeping in tune with conditions. A master knows who is active, from where, and when. Make operating your radio station a natural and comfortable activity by keeping yourself in shape with regular radio exercise.

Paying Attention to Detail

Masters know that the little things are what make the difference between 100-percent and 90-percent performance or even between being on and off the air. Waterproofing that connector completely or having your CQ sound just right really pays off in the long run. And masters are on the radio for the long run.

The Problem Ain't What You Don't Know

It's what you do know that ain't so! (Will Rogers) If you guess wrong, don't be too proud to admit it. Learn the right way or correct fact. The worst mistakes are made by ignoring the truth. A master isn't afraid to say those three dreaded words, "I don't know!"

Antennas Make the Difference

If you look at the top stations in any facet of ham radio, you'll find that their owners spend the most time and effort on the antenna systems. You'll find no better return on investment in ham radio than in improving your antennas. Masters are often antenna and feedline gurus.

A Decibel Is a Decibel Is a Decibel

Masters know that any improvement in the path between operators is not to be discounted. Anything that makes your signal easier to understand — one decibel less noise received, one decibel better audio quality, one decibel stronger transmitted signal — makes holding the contact easier.

Ham Radio Is a Lifetime of Learning

Take advantage of every learning opportunity, including learning from your mistakes . . . you'll have plenty! A master knows that each problem or goof is also a lesson. Masters get to be masters by starting as raw recruits just like you and then making one improvement at a time, day in and day out.

Chapter 18

Ten First Station Tips

In This Chapter

▶ Avoiding mistakes in the creation of your first station

▶ Saving money on equipment

*W*hen you're putting your first station together — whether at home, mobile, portable, or even just a hand-held radio — getting sidetracked and creating problems later is easy. Avoid common pitfalls by applying the simple tips I give in this chapter. That way, you won't scratch your head later, thinking, "Why did I do that?"

Being Flexible

Don't assume that you'll be doing the same activities on the air forever. Avoid over-specialized gear except where required for a specific type of operating. Make use of the flexibility of a computer and software to implement functions that are likely to change, such as a digital signal. Don't nail everything down: Allow equipment to be moved around for comfort and layout convenience. The built-in look is attractive, but very hard to change later.

Looking and Learning

Browse the Web and read articles that show how other stations are put together. Make note of any particularly good ideas. Don't hesitate to write or e-mail the station owners to ask questions — they welcome your attention and interest. Take advantage of opportunities to visit local stations, too.

Don't Put All Your Eggs in One Basket

Spending a lot of money on a radio right away is tempting, but you'll find yourself needing other gear, such as antennas and cables, that you perhaps hadn't counted on. Those extras can add up to at least as much as your main radio, so leave yourself some budget for them, too.

Used-Equipment Bargains

If you have a knowledgeable friend who can help you separate the wheat from the chaff, used equipment is a great way to get started. By saving money, you have more cash for exploring new modes and bands later on. It's *caveat emptor,* though: You can easily encounter junk equipment. If in doubt, or if the deal seems too good to be true, pass it up.

Building Something!

Using equipment you build yourself is a thrill. Your skills benefit from small construction projects such as audio switches, filters, or keyers. Buying everything new is generally less trouble, but the equipment is expensive, too. Building some things yourself can save you some money. Don't be afraid to get out the drill and soldering iron. You can find lots of kits, magazines, and handbook articles to get you started.

Being Well-Grounded

Don't neglect grounding. Put in a ground system (see Chapter 13) as the first step — adding grounding after the equipment and wires are already in place is much harder. Good grounding helps you avoid RF feedback and ground loops later, both frustrating and aggravating problems.

Saving Money by Building Your Own Cables

You need lots of cables and connectors in your station. At a cost of roughly $5 or more for each pre-made cable, you can quickly spend as much on connecting your equipment as you can on a major accessory. Learn how to install connectors well and you save many, many dollars over the course of your ham career.

Building Step-by-Step

After you have the basics of your shack in place, you can upgrade your equipment in steps so that you can always hear a little farther than you can transmit. Don't be an *alligator* (all mouth, no ears)! Plan ahead with a goal in mind so that your ham radio dollars and hours all work to further that goal.

Finding the Weakest Link

Every station has a weakest link. Always be on the lookout for a probable point of failure or of loss of quality. On the airwaves, you'll encounter stations with muffled or distorted audio that have a multi-kilobuck radio but a cheap, garage-sale microphone. Use quality gear and keep heavily-used equipment well-maintained.

Being Comfortable

You're going to spend a lot of hours in front of your radio so take care of the operator, too. Start with a comfortable chair — excellent chairs are often available in used office furniture stores at far below new cost. Make sure you have adequate lighting and that the operating desk is at a comfortable height. Those dollars will pay dividends every time you sit down!

Chapter 19

Ten Easy Ways to Have Fun on the Radio

So you're sitting there saying, "There's nothing to do!" Don't worry! Everybody gets in a bit of a rut now and then. In this chapter, I give you ten great ideas to shake off the radio blues and spice up your operating.

Listening for People Having Fun and Joining In

Sounds simple, doesn't it? Just turn on the radio and start tuning the bands. Join a ragchew, check into a net, monitor a slow-scan transmission, listen for someone calling "CQ Contest," or find a pileup and dive in. On any given day, literally hundreds of different activities are taking place and all you need to do to participate is to just spin the tuning dial and pay attention.

Special Events and Contests Are Looking for You!

Every weekend, you can find little pockets of activity sprinkled around the HF and VHF bands as contests and special event stations take to the airwaves. Nearly every contest is open to the casual passer-by. The operators involved welcome your call and help you exchange the necessary information. You can contact special event stations and often have beautiful, interesting certificates for contacts.

Making Up Your Own Contest

Get together with your friends or club and dream up a silly competition just for fun. Buy an old bowling trophy and make up a challenge to go with it. See who can contact the most states in a weekend. Play radio bingo. Tune up and down the band with a friend and see who can make contacts with the lowest power. Go nuts, it's a hobby!

Sending a Radiogram, Ma'am

Who is on your "I should write" list? Maybe a relative or friend would enjoy getting a short radiogram for a birthday or just to say hello. Look up the NTS local nets in your neighborhood, fill out the radiogram form, and jump in. ***Warning:*** It's addictive!

Joining the Parade

Every town has public events such as parades, festivals, sports, and concerts that make use of hams to assist them with communications and other electronics chores. You'll feel great after giving these folks a hand and they will have an appreciation of ham radio, too. Contact your ARRL Section Manager for information about how to get in touch with organizers or ham public service groups.

Going Somewhere Cool

For a real treat, try mobile or portable operation from some unusual place. On HF, try operating from a rural county or county line on the County Hunters Net. On VHF and UHF, with a short drive, you can be in a sought-after grid square. The nearest hilltop or scenic overlook can generate hours of fun. Take a photo and make up a neat QSL card to go with the contacts!

Squirting a Bird

Making contacts through a satellite (*squirting a bird*) is a lot easier than you think. Try the FM repeater or packet satellites for starters. Monitor the International Space Station and Space Shuttle downlink frequencies. You'll receive an interesting certificate just for reporting that you heard either one. Watching their reflections move across the sky at dusk or dawn will never be the same after you make contact with them.

Learning a New Lingo

Brush up those high-school foreign language skills and make contact with a DX station. If you're equipped for HF, you can talk to them directly. Chapter 9 explains how to make VHF/UHF contacts via repeaters using IrLP or one of the other linking systems. Ham radio makes learning a new language easy. Just learn a few new words at a time, and soon you have a whole new vocabulary and a DX friend or two, to boot.

Shortwave Listening (SWL-ing)

What's on the frequencies between the HF ham bands? Most radios now have general-coverage receivers that tune all frequencies, so why not tune in the German national broadcaster Deutsche Welle or HCJB from Quito, Ecuador? Check the Web sites or get a copy of a shortwave listener's guide to find English language programs. And, of course, music is always available and requires no translation to enjoy.

Visiting a New Group

Drop in on a new club or net any time. They'll be pleased you are visiting and you may stumble into a whole new outlook on the hobby. I particularly enjoy finding a club program on a subject I'm not acquainted with and making a friend or two. Be sure to invite them to your meetings, as well!

Chapter 20

Ten Ways to Give Back to Ham Radio

. . . .

In This Chapter

▶ Tips for contributing to the ham radio community

▶ Preparing for emergencies

▶ Becoming an Elmer

. . . .

*H*am radio provides you with so many wonderful things to do and learn, yet it is a service. The expectation is that beyond using ham radio just for personal enjoyment, you contribute a little back to the public for use of the airwaves. I show you ten ways to make a contribution in this chapter.

Preparing Yourself for Emergencies

The best thing you can do for emergency preparedness is to be sure you and your family are ready. In an emergency, take care of home and family first. Only then should you think of ham radio. If you're not ready at home, you can't provide assistance to others.

Preparing Your Community for Emergencies

Find out what emergency groups are active in your community. The ARRL Section Manager or District Emergency Coordinator is a good place to start asking. If a group is active, join it. If no group exists, maybe you should consider starting one.

Volunteering in Your Club

All clubs need individuals willing to put their shoulders to the wheel. Whether you are limited to minor services or can volunteer for a leadership position, your time and effort is welcomed. If you are a new member, you'll find no better way to become part of the family than to help out with a chore, no matter how small.

Performing Public Service Assistance

Public service assistance is an easy and rewarding way to make a contribution. Start by asking the ARRL Section Manager if you can pitch in at a fun run, parade, or sporting event. By lending a hand, you help everything run more smoothly and learn a lot about emergency communications.

Experimenting

The Amateur Service is also intended to foster technical innovation in radio technology and techniques. You don't have to be a Nobel laureate to try out a new antenna design, write a simple program, or experiment with propagation. Also, telling others about your results is easier than you may think. Newsletters and magazines love to help you spread the word.

Participating in On-the-Air Monitoring

The amateur bands are often hosts to unwanted intruders who take advantage of the bands to avoid licensing fees or tests. The ARRL's Intruder Watch program needs your ears to help keep these freeloaders out. If you'd like to assist ham radio's self-policing nature, consider becoming an Official Observer. NOAA's SKYWARN program and numerous local weather nets depend on the contributions of hams, as well.

Acting as a Product Tester or QSL Manager

Hardware manufacturers and software authors often need testing assistance from hams, the potential users of their products. Keep a watch for requests for testers on Web sites and in Internet forums. Also, DX stations often need help with replying to the many QSL requests they receive. Offering your services as a manager furthers the sport of DX-ing and international goodwill.

Representing Amateur Radio

Write your government representatives on issues affecting ham radio. Don't limit yourself to state and federal issues. Many things happen on the local level, such as zoning, planning, covenants, and permitting that could use your input. Conversely, discovering planned actions that have an adverse affect on the service can save us all a lot of trouble, if modified in time.

Being an Elmer

You're a full-fledged ham now, so you can put on the Elmer hat and give a hand to others just starting out. Being new to the hobby yourself, you are in a great position to understand what a newcomer needs to know and what seems confusing.

Making Lifelong Friendships

I would be omitting one of the most important aspects of building and maintaining a vital and dynamic service if I did not mention the friendships that hams form. They knit the hobby together and make it an enjoyable activity to return to over the course of a lifetime. By enjoying each other's company and sharing in successes and failures, we build a community that grows stronger with every new voice.

Part VI
Appendixes

The 5th Wave   By Rich Tennant

"Saaay — I have an idea. Why don't we turn down the lights, put on some soft music, and practice our Morse code."

In this part . . .

I include two appendixes for use as handy references. Appendix A includes a glossary of all the terms you come across in *Ham Radio For Dummies*.

Appendix B is a collection of great books, articles, and Web sites that you can use. You may want to add them to your browser bookmarks list for even easier reference. The reference texts make a great birthday or holiday gift, don't you think?

Appendix A

Glossary

• •

*T*his glossary is reprinted with the permission of the ARRL.

Amateur operator: A person holding a written authorization to be the control operator of an amateur station.

Amateur service: A radio communication service for the purpose of self-training, intercommunication, and technical investigations carried out by amateurs, that is, duly authorized persons interested in radio technique solely with a personal aim and without pecuniary interest.

Amateur station: A station licensed in the amateur service, including necessary equipment, used for amateur communication.

Ammeter: A test instrument that measures current.

Ampere (A): The basic unit of electrical current. Current is a measure of the electron flow through a circuit. If you count electrons, 6.24×10^{18} electrons moving past a point in one second equal a current of one ampere. Abbreviated as *amps.* (Numbers written as a multiple of some power are expressed in exponential notation, as shown here.)

Amplitude modulation (AM): A method of combining an information signal and an RF (radio-frequency) carrier. In voice AM transmission, the voice information can vary (modulate) the amplitude of an RF carrier. Shortwave broadcast stations use this type of AM, as do stations in the Standard Broadcast Band (535 to 1710 kHz). A variation of AM, known as single sideband, is very popular.

Antenna: A device that picks up or sends out radio frequency energy.

Antenna switch: A switch that connects one transmitter, receiver, or transceiver to several different antennas.

Antenna tuner: A device that matches the antenna system input impedance to the transmitter, receiver, or transceiver output impedance. Also called an *antenna-matching network, impedance-matching network,* or *Transmatch.*

Autopatch: A device that allows repeater users to make telephone calls through a repeater.

Balun: Contraction for balanced to unbalanced. A device to couple a balanced load to an unbalanced source, or vice versa.

Band spread: A receiver quality that describes how far apart stations on different nearby frequencies seem to be. Usually expressed as the number of kilohertz that the frequency changes per tuning-knob rotation. The amount of band spread determines how easily signals can be tuned.

Band-pass filter: A circuit that allows signals to go through it only if the signals are within a certain range of frequencies. It attenuates signals above and below this range.

Bandwidth: The width of a frequency band outside of which the mean power is attenuated at least 26dB below the mean power of the total emission, including allowances for transmitter drift or Doppler shift. Bandwidth describes the range of frequencies that a radio transmission occupies.

Battery: A device that converts chemical energy into electrical energy.

Beacon station: An amateur station transmitting communications for the purposes of observation of propagation and reception or other related experimental activities.

Beam antenna: A directional antenna. A beam antenna must be rotated to provide its strongest coverage in different directions.

Beat-frequency oscillator (BFO): A receiver circuit that provides a signal to the detector. The BFO signal mixes with the incoming signal to produce an audio tone for CW reception. A BFO is needed to copy CW and SSB signals.

Broadcasting: Transmissions intended to be received by the general public, either direct or relayed.

Capacitor: An electrical component usually formed by separating two conductive plates with an insulating material. A capacitor stores energy in an electric field.

Chirp: A slight shift in transmitter frequency each time you key the transmitter.

Closed repeater: A repeater that restricts access to those who know a special code.

Coaxial cable (Coax): Pronounced *kó-aks*. A type of feedline with one conductor inside the other.

Continuous wave (CW): Morse code telegraphy.

Control operator: An amateur operator designated by the licensee of a station to be responsible for the transmissions of an amateur station.

Control point: The locations at which the control operator functions are performed.

Courtesy tone: A tone or beep transmitted by a repeater to indicate that the next station can begin transmitting. The courtesy tone is designed to allow a pause between transmissions on a repeater, so other stations can call. It also indicates that the time-out timer has been reset.

CQ: The general call when requesting a conversation with anyone — "Calling any station."

Crystal oscillator: A device that uses a quartz crystal to keep the frequency of a transmitter constant.

Crystal-controlled transmitter: A simple type of transmitter that consists of a crystal oscillator followed by driver and power amplifier stages.

CTCSS (Continuous Tone Coded Squelch System): A sub-audible tone system used on some repeaters. When added to a carrier, a CTCSS tone allows a receiver to accept a signal. Also called *PL.*

Cubical quad antenna: An antenna built with its elements in the shape of four-sided loops.

Current: A flow of electrons in an electrical circuit.

CW (Morse code): A communications mode transmitted by on/off keying of a radio-frequency signal. Another name for international Morse code.

D region: The lowest region of the ionosphere. The D region contributes very little to shortwave radio propagation. It acts mainly to absorb energy from radio waves as they pass through it. This absorption has a significant effect on signals below about 7.5 MHz during daylight.

Data: Computer-based communications modes, such as packet radio, which can be used to transmit and receive computer files, or digital information.

DE: The Morse code abbreviation for "from" or "this is."

Delta loop antenna: A variation of the cubical quad with triangular elements.

Digipeater: A packet-radio station used to retransmit signals that are specifically addressed to be retransmitted by that station.

Digital communications: Computer-based communications modes. These modes can include data modes, such as packet radio, and text-only modes like radioteletype (RTTY).

Dipole antenna: *See* ½-wave dipole. A dipole not ½ wavelength long is called a "doublet."

Director: An element in front of the driven element in a Yagi antenna and some other directional antennas.

Driven element: The part of an antenna that connects directly to the feedline.

Dual-band antenna: An antenna designed for use on two different Amateur Radio bands.

Dummy antenna: A station accessory that allows you to test or adjust transmitting equipment without sending a signal out over the air. Also called *dummy load.*

Dummy load: *See* Dummy antenna.

Duplexer: A device that allows a dual-band radio to use a single dual-band antenna.

Duty cycle: A measure of the amount of time a transmitter is operating at full output power during a single transmission. A lower duty cycle means less RF radiation exposure for the same PEP output.

DX: Distant, foreign countries.

E region: The second lowest ionospheric region, the E region exists only during the day. Under certain conditions, it may refract radio waves enough to return them to Earth.

Earth ground: A circuit connection to a ground rod driven into the Earth or to a cold-water pipe made of copper that goes into the ground.

Earth station: An amateur station located on, or within 50 kilometers of, the Earth's surface intended for communications with space stations or with other Earth stations by means of one or more other objects in space.

Earth-Moon-Earth (EME) or Moonbounce: A method of communicating with other stations by reflecting radio signals off the moon's surface.

Electron: A tiny, negatively charged particle, normally found in an area surrounding the nucleus of an atom. Moving electrons make up an electrical current.

Emergency traffic: Messages with life and death urgency or requests for medical help and supplies that leave an area shortly after an emergency.

Emission: The transmitted signal from an amateur station.

Emission privilege: The permission granted by your license to use a particular emission type (such as Morse code or voice).

Emission types: Term for the different modes authorized for use on the Amateur Radio bands. Examples are CW, SSB, RTTY, and FM.

F region: A combination of the two highest ionospheric regions, the F1 and F2 regions. The F region refracts radio waves and returns them to Earth. Its height varies greatly depending on the time of day, season of the year, and amount of sunspot activity.

False or deceptive signals: Transmissions intended to mislead or confuse those who may receive the transmissions. For example, transmitting distress calls with no actual emergency are false or deceptive signals.

Feedline: The wires or cable used to connect a transmitter, receiver, or transceiver to an antenna. *See* Transmission line.

Filter: A circuit that allows some signals to pass through it but greatly reduces the strength of others.

"Five-Nine": A common signal report on voice that means "Your signal is strong and easy to understand." The equivalent on CW or Morse code is "599." *See also* Signal Report.

Frequency: The number of complete cycles of an alternating current that occur per second.

Frequency bands: A group of frequencies where amateur communications are authorized.

Frequency coordination: Allocating repeater input and output frequencies to minimize interference between repeaters and to other users of the band.

Frequency coordinator: An individual or group that recommends repeater frequencies to reduce or eliminate interference between repeaters operating on or near the same frequency in the same geographical area.

Frequency discriminator: A type of detector used in some FM receivers.

Frequency modulated (FM) phone: The type of signals used to communicate by voice (phone) over most repeaters. FM is a method of combining an RF carrier with an information signal, such as voice. The voice information (or data) changes the RF carrier frequency in the modulation process (*see* Amplitude modulation). Voice or data vary the frequency of the transmitted signal. FM broadcast stations and most professional communications (police, fire, taxi) use FM. VHF/UHF FM voice is the most popular amateur mode.

Frequency privilege: The permission granted by your license to use a particular group of frequencies.

Front-end overload: Interference to a receiver caused by a strong signal that overpowers the receiver RF amplifier (front end). *See also* receiver overload.

General-coverage receiver: A receiver used to listen to a wide range of frequencies. Most general-coverage receivers tune from frequencies below the standard-broadcast band to at least 30 MHz. These frequencies include the shortwave-broadcast bands and the amateur bands from 160 to 10 meters.

Grace period: The time the FCC allows following the expiration of an amateur license to renew the license without having to retake an examination. Hams holding an expired license may not operate an amateur station until the license is reinstated.

Ground connection: A connection made to the Earth for electrical safety. You can make this connection inside (to a metal cold-water pipe) or outside (to a ground rod).

Ground rod: A copper or copper-clad steel rod driven into the Earth. A heavy copper wire from the ham shack connects all station equipment to the ground rod.

Ground-wave propagation: The method by which radio waves travel along the Earth's surface.

Half-wave dipole: A basic antenna used by radio amateurs. It consists of a length of wire or tubing, opened and fed at the center. The entire antenna is ½ wavelength long at the desired operating frequency.

Ham-bands-only receiver: A receiver designed to cover only the bands used by amateurs. Usually refers to the bands from 80 to 10 meters, sometimes including 160 meters.

Harmonics: Signals from a transmitter or oscillator occurring on whole-number multiples (2_, 3_, 4_) of the desired operating frequency.

Health and Welfare traffic: Messages about the well being of individuals in a disaster area. Such messages must wait for Emergency and Priority traffic to clear, and results in advisories to those outside the disaster area awaiting news from family and friends.

Hertz (Hz): An alternating-current frequency of one cycle per second. The basic unit of frequency.

High-pass filter: A filter designed to pass high-frequency signals, while blocking lower-frequency signals.

Impedance-matching device: *See* Antenna Tuner.

Input frequency: A repeater's receiving frequency. To use a repeater, transmit on the input frequency and receive on the output frequency.

Intermediate frequency (IF): The output frequency of a mixing stage in a superheterodyne receiver. The subsequent stages in the receiver are tuned for maximum efficiency at the IF.

Ionizing radiation: Electromagnetic radiation that has sufficient energy to knock electrons free from their atoms, producing positive and negative ions. X-rays, gamma rays, and ultraviolet radiation are examples of ionizing radiation.

Ionosphere: A region of electrically charged (ionized) gases high in the atmosphere. The ionosphere bends radio waves as they travel through it, returning them to Earth. *See also* Sky-wave Propagation.

K: The Morse code abbreviation for "any station respond."

Lightning protection: You can help prevent lightning damage to your equipment (and your house) in several ways, among them unplugging equipment, disconnecting antenna feedlines, and using a lightning arrestor.

Limiter: A stage of an FM receiver that makes the receiver less sensitive to amplitude variations and pulse noise.

Line-of-sight propagation: The term used to describe VHF and UHF propagation in a straight line directly from one station to another.

Lower sideband (LSB): The common single-sideband operating mode on the 40, 80, and 160-meter amateur bands.

Low-pass filter: A filter that allows signals below the cutoff frequency to pass through and attenuates signals above the cutoff frequency.

Malicious (harmful) interference: Intentional, deliberate obstruction of radio transmissions.

Maximum useable frequency (MUF): The highest-frequency radio signal that reaches a particular destination using sky-wave propagation, or skip. The MUF may vary for radio signals sent to different destinations.

MAYDAY: From the French *m'aidez* (help me), MAYDAY is used when calling for emergency assistance in voice modes.

Microphone: A device that converts sound waves into electrical energy.

Mobile device: A radio transmitting device that you can mount in a vehicle. A push-to-talk (PTT) switch activates the transmitter.

Modem: Short for *modulator/demodulator*. A modem modulates a radio signal to transmit data and demodulates a received signal to recover transmitted data.

Modulate: To vary the amplitude, frequency, or phase of a radio-frequency signal.

Modulation: The process of varying an RF carrier in some way (the amplitude or the frequency, for example) to add an information signal to be transmitted.

Monitor mode: One type of packet radio receiving mode. In monitor mode, everything transmitted on a packet frequency is displayed by the monitoring TNC. The data is displayed whether or not the transmissions are addressed to the monitoring station.

Morse code: *See* CW.

Multimode transceiver: Transceiver capable of SSB, CW, and FM operation.

National Electrical Code: A set of guidelines governing electrical safety, including antennas.

Network: A term used to describe several packet stations linked together to transmit data over long distances.

Nonionizing radiation: Electromagnetic radiation that does not have sufficient energy to knock electrons free from their atoms. Radio frequency (RF) radiation is nonionizing.

Offset: For CW operation, the 300 to 1000-Hz difference in transmitting and receiving frequencies in a transceiver. For a repeater, offset refers to the difference between its transmitting and receiving frequencies.

One-way communications: Transmissions not intended to be answered. The FCC strictly limits the types of one-way communications allowed on the amateur bands.

Open repeater: A repeater used by all hams who have a license that authorizes operation on the repeater frequencies.

Operator/primary station license: An amateur license actually includes two licenses in one. The operator license is that portion of an Amateur Radio license that gives permission to operate an amateur station. The primary station license is that portion of an Amateur Radio license that authorizes an amateur station at a specific location. The station license also lists the call sign of that station.

Output frequency: A repeater's transmitting frequency. To use a repeater, transmit on the input frequency and receive on the output frequency.

Packet radio: A system of digital communication whereby information is broken into short bursts. The bursts *(packets)* also contain addressing and error-detection information.

Parasitic beam antenna: *See* Beam Antenna.

Parasitic element: Part of a directive antenna that derives energy from mutual coupling with the driven element. Parasitic elements are not connected directly to the feedline.

Peak envelope power (PEP): The average power of a signal at its largest amplitude peak.

Pecuniary: Payment of any type, whether money or other goods. Amateurs may not operate their stations in return for any pecuniary.

Phone: Another name for voice communications.

Phone emission: The FCC name for voice or other sound transmissions.

Phonetic alphabet: Standard words used on voice modes, which make understanding letters of the alphabet easier, such as those in call signs. The call sign KA6LMN stated phonetically is Kilo Alfa Six Lima Mike November.

PL: *See* CTCSS.

Polarization: The electrical-field characteristic of a radio wave. An antenna parallel to the surface of the Earth, such as a dipole, produces horizontally polarized waves. An antenna perpendicular to the Earth's surface, such as a ¼-wave vertical, produces vertically polarized waves. An antenna with both horizontal and vertical polarization is circularly polarized.

Portable device: A radio transmitting device designed to have a transmitting antenna that is generally within 20 centimeters of a human body.

Priority traffic: Emergency-related messages, but not as important as Emergency traffic.

Procedural signal (prosign): One or two letters sent as a single character. Amateurs use prosigns in CW contacts as a short way to indicate the operator's intention. Some examples are K for "Go Ahead," or AR for "End of Message."

Product detector: A device that allows a receiver to process CW and SSB signals.

Propagation: The study of how radio waves travel.

Q signals: Three-letter symbols beginning with Q. Used on CW to save time and to improve communication. Some examples are QRS (send slower), QTH (location), QSO (ham conversation), and QSL (acknowledgment of receipt).

QRL?: Ham radio Q signal meaning "Is this frequency in use?"

QRP: Ham radio Q signal meaning "Low Power." QRP generally means to 5 watts of transmitted power on CW or 10 watts of peak power on phone. QRPp means power less than 1 watt.

QSL card: A postcard that serves as a confirmation of communication between two hams.

QSO: A conversation between two radio amateurs.

Quarter-wavelength vertical antenna: An antenna constructed of a ¼-wavelength long radiating element placed perpendicular to the Earth.

Radio Amateur Civil Emergency Service (RACES): A part of the Amateur Service that provides radio communications for civil preparedness organizations during local, regional, or national civil emergencies.

Radio-frequency interference (RFI): Disturbance to electronic equipment caused by radio-frequency signals.

Radioteletype (RTTY): Radio signals sent from one teleprinter machine to another machine. Anything that one operator types on his teleprinter prints on the other machine.

Receiver: A device that converts radio waves into signals you can hear or see.

Receiver incremental tuning (RIT): A transceiver control that allows for a slight change in the receiver frequency without changing the transmitter frequency. Some manufacturers call this control a *clarifier (CLAR) control.*

Receiver overload: Interference to a receiver caused by a strong RF signal that forces its way into the equipment. A signal that overloads the receiver RF amplifier (front end) causes front-end overload. Receiver overload is sometimes called *RF overload.*

Reflection: Signals that travel by line-of-sight propagation are reflected by large objects, such as buildings.

Reflector: An element behind the driven element in a Yagi antenna and other directional antennas.

Repeater station: An amateur station that automatically retransmits the signals of other stations.

RF burn: A burn produced by coming in contact with exposed RF voltages.

RF carrier: A steady radio frequency signal that is modulated to add an information signal to be transmitted. For example, a voice signal is added to the RF carrier to produce a phone emission signal.

RF overload: Another term for receiver overload.

RF radiation: Waves of electric and magnetic energy. Such electromagnetic radiation with frequencies as low as 3 kHz and as high as 300 GHz are considered part of the RF region.

RF safety: Preventing injury or illness to humans from the effects of radio-frequency energy.

Rig: The radio amateur's term for a transmitter, receiver, or transceiver.

RST: A system of numbers used for signal reports: R is readability, S is strength, and T is tone. (On single-sideband phones, only R and S reports are used.)

Selectivity: The ability of a receiver to separate two closely spaced signals.

Sensitivity: The ability of a receiver to detect weak signals.

73: Ham lingo for "best regards." Used on both phone and CW toward the end of a contact.

Shack: The room where an Amateur Radio operator keeps his or her station equipment.

Sidebands: The sum or difference frequencies generated when an RF carrier mixes with an audio signal. Single-sideband phone (SSB) signals have an upper sideband (USB) and a lower sideband (LSB). SSB transceivers allow operation on either USB or LSB. *See also* USB and LSB.

Signal report: A set of numbers that are exchanged to indicate the relative quality of a signal's quality in terms of strength, clarity, and purity (*see* RST).

Simplex operation: Receiving and transmitting on the same frequency.

Single Sideband (SSB) phone: A common mode of voice operation on the amateur bands. SSB is a form of amplitude modulation. The amplitude of the transmitted signal varies with the voice signal variations.

Skip zone: An area of poor radio communication that is too distant for ground waves and too close for sky waves.

Sky-wave propagation: The method radio waves travel through the ionosphere and back to Earth. Sometimes called *skip,* sky-wave propagation has a far greater range than line-of-sight and ground-wave propagation.

SOS: A Morse code call for emergency assistance.

Space station: An amateur station located more than 50 km above the Earth's surface.

Specific absorption rate (SAR): A term that describes the rate RF energy is absorbed into the human body. Maximum permissible exposure (MPE) limits are based on whole-body SAR values.

Splatter: A type of interference to stations on nearby frequencies. Splatter occurs when a transmitter is overmodulated.

Spurious emissions: Signals from a transmitter on frequencies other than the operating frequency.

Standing-wave ratio (SWR): Sometimes called voltage standing-wave ratio (VSWR). A measure of the impedance match between the feedline and the antenna. Also, with a Transmatch in use, a measure of the match between the feedline from the transmitter and the antenna system. The system includes the Transmatch and the line to the antenna. VSWR is the ratio of maximum voltage to minimum voltage along the feedline. Also the ratio of antenna impedance to feedline impedance when the antenna is a purely resistive load.

Station grounding: Connecting all station equipment to a good Earth ground improves both safety and station performance.

Sunspot cycle: The number of sunspots increases and decreases in a predictable cycle that lasts about 11 years.

Sunspots: Dark spots on the surface of the sun. With a few sunspots, long-distance radio propagation is poor on the higher-frequency bands. With many sunspots, long-distance HF propagation improves.

SWR meter: A measuring instrument that indicates when an antenna system is working well. A device used to measure SWR (*see* Standing-wave ratio).

Tactical call signs: Names used to identify a location or function during local emergency communications.

Teleprinter: A machine that can convert keystrokes (typing) into electrical impulses. The teleprinter also converts the proper electrical impulses back into text. Computers have largely replaced teleprinters for amateur radiotele-type work.

Television interference (TVI): Interruption of television reception caused by another signal.

Temperature inversion: A condition in the atmosphere in which a region of cool air is trapped beneath warmer air.

Temporary state of communications emergency: When a disaster disrupts normal communications in a particular area, the FCC can declare this type of emergency. Certain rules may apply for the duration of the emergency.

Terminal: An inexpensive piece of equipment used in place of a computer in a packet radio station.

Third-party communications: Messages passed from one amateur to another on behalf of a third person.

Third-party communications agreement: An official understanding between the United States and another country that allows amateurs in both countries to participate in third-party communications.

Third-party participation: The way an unlicensed person can participate in amateur communications. A control operator must ensure compliance with FCC rules.

Ticket: A common name for an Amateur Radio license.

Time-out timer: A device that limits the amount of time any one person can talk through a repeater.

Transceiver: A radio transmitter and receiver combined in one unit.

Transmission line: The wires or cable used to connect a transmitter or receiver to an antenna. Also called a *feedline*.

Transmitter: A device that produces radio-frequency signals.

Troposphere: The region in the Earth's atmosphere just above the Earth's surface and below the ionosphere.

Tropospheric bending: When radio waves are bent in the troposphere, they return to Earth farther away than the visible horizon.

Tropospheric ducting: A type of VHF propagation that occurs when warm air overruns cold air (a temperature inversion).

Unbalanced line: A feedline with one conductor at ground potential, such as a coaxial cable.

Uncontrolled environment: Any area in which an RF signal may cause radiation exposure to people who may not be aware of the radiated electric and magnetic fields. The FCC generally considers members of the general public and an amateur's neighbors to be in an uncontrolled RF radiation exposure environment to determine the maximum permissible exposure levels.

Unidentified communications or signals: Signals or radio communications in which the transmitting station's call sign is not transmitted.

Upper sideband (USB): The common single-sideband operating mode on the 20, 17, 15, 12, and 10-meter HF amateur bands, and all the VHF and UHF bands.

Vertical antenna: A common amateur antenna, often made of metal tubing. The radiating element is vertical. Usually four or more radial elements are parallel to or on the ground.

VFO: Variable Frequency Oscillator — the circuit in a receiver or transmitter that controls the operating frequency.

Visible horizon: The most distant point you see by line of sight.

Voice: Any of the several methods used by amateurs to transmit speech.

Voice communications: Hams can use several voice modes, including FM and SSB.

Wavelength: Often abbreviated l. The distance a radio wave travels in one RF cycle. The wavelength relates to frequency. Higher frequencies have shorter wavelengths.

Yagi antenna: The most popular type of amateur directional (beam) antenna. It has one driven element and one or more additional elements.

Appendix B

The Best References

This appendix is a listing of many useful Web sites and books that can help answer your many questions as you start your ham radio career. Most ham radio books are available from the ARRL at www.arrl.org/catalog. Where a book has a special publisher or other source, it's noted with a Web address where you can find it.

Web Portals

The Web portals referenced here are good "newsstands" for ham radio. On these sites, you'll find information about current and upcoming events, radio conditions, and news stories. These sites also host mailing lists and forums on various topics, equipment swap 'n shops, archived files and photos, and product reviews. They are designed to be your ham radio home page.

eHam.net

This portal features news and articles, e-mail discussion groups, product reviews, for sale listings, DX spotting and solar information, and surveys, among other things.

www.eham.net

QRZ.com

QRZ.com is a general interest portal and call sign/licensee lookup facility that features extensive articles, news and discussion groups, and online practice licensing exams.

```
www.qrz.com
```

K3TKS' QSL.net

QSL.net is host to hundreds of individual and club ham radio Web pages and e-mail reflectors. You can search the links and pages.

```
www.qsl.net
```

AC6V's Amateur Radio and DX Reference Guide

This site offers many, many links and references covering all phases of ham radio.

```
www.ac6v.com
```

Yahoo! Amateur and Ham Radio Directory

Here you can find links to a large number of general purpose and specialized Web sites.

```
dir.yahoo.com/News_and_Media/Radio/Amateur_and_Ham_Radio
```

Buckmaster Publishing Hamcall

This site provides a worldwide call sign lookup service.

```
hamcall.net/call
```

Amateur Radio Webring

The Amateur Radio Webring is a list of many ham radio Web sites for clubs and resource pages.

```
g.webring.com/hub?ring=amateurradio
```

Operating References

No one knows or remembers everything, so it's a good idea to have references on hand to guide your on-the-air activities. Here are some books you'll find handy on a day-to-day basis while operating.

The FCC Rule Book

The rule book, published by the ARRL, explains in clear text what the regulations mean and how you apply them. It also includes the actual text of the Part 97 rules.

The ARRL Operating Manual

The operating guide, published by the ARRL, covers nearly all phases of ham radio operating, including maps and numerous references.

On the Air with Ham Radio, by Steve Ford, WB8IMY

This book is a good introductory text to help you set up a station and get on the air.

The ARRL Repeater Directory

Published by the ARRL, this directory lists North American repeaters on 10-meters through the UHF and microwave bands.

Public service

After you start performing public service activities, you'll need forms and training. Luckily, most of what you need is available online at the following sites.

ARRL Public Service Web page

This site includes several operating manuals available for free download, guidelines and brochures, and links to other emergency communications organizations.

```
www.arrl.org/FandES/field/pubservice.html
```

Amateur Radio Emergency Communications Courses: Level 1, 11, 111

These online courses introduce the ham to "emcomm," net control and management, and team management and planning. You may be reimbursed the course tuition based on grant availability.

```
www.remote.arrl.org/cce/courses.html
```

Net Directory Search

You can use the online Net Directory Search to find on-the-air nets by frequency, name and topic, or region.

```
www.arrl.org/FandES/field/nets/client/netsearch.html
```

Digital modes

The online information available at these sites really helps you get going on the digital modes. TAPR offers broad coverage of many different modes and protocols. K1VY's site focuses on PSK31, the most popular of the new digital modes.

Tucson Amateur Packet Radio (TAPR)

The TAPR is the biggest ham group specializing in digital modes. Its Web site is a smorgasbord of information about all the popular digital protocols.

```
www.tapr.org
```

PSK31

Neil Rosenberg K1VY put together a comprehensive site on PSK31. You find quite a few descriptive articles and construction projects, as well as links to other digital data sites.

```
psk31.com
```

DX-ing resources

Successful DX-ing takes timely information about which stations are active and what conditions you can expect on the air. The following references are DX-ing magazines, newsletters, and bulletins containing information about currently active and upcoming DX events:

- ✔ *DX Magazine* (`www.dxpub.com/dx_mag.html`): This bimonthly periodical contains articles about the techniques of DX-ing and includes numerous travelogues of "DX-peditions" to interesting places around the world.

- ✔ *QRZ DX* (`www.dxpub.com/qrz_dx_nl.html`): *QRZ DX* is a weekly print or e-mail newsletter with DX-ing news, a substantial listing of IOTA activity (island-based activity), and a listing of frequencies on which sought-after stations have been heard.

- ✔ *Daily DX* (`www.dailydx.com`): *Daily DX* covers news and activity as a daily e-mail newsletter full of late-breaking news and on-the-air surprises.

- ✔ **The *OPDX Bulletin*** (`www.papays.com/opdx.html`) **and *ARRL DX Bulletin*** (`www.arrl.org/w1aw/dx`)**:** These are free weekly e-mail bulletins containing a compendium of current and anticipated DX station activities.

The Complete DXer, by Bob Locher W9KNI

This book should be read by every budding DXer to learn the basics of DX-ing and good operating. In a wonderfully readable short-story format, Bob explains the right ways to go about putting DX call signs in your log. If you're mystified by what you hear while chasing DX, W9KNI has been there and helps you understand the whys and hows. The book is available from Idiom Press at www.idiompress.com.

DXsummit.com

DXsummit.com is a worldwide DX spotting Web site where hams post reports on every band from every land 24 hours a day. The reports, known as *spots,* are searchable by call and band. The site also records solar data for propagation information.

```
www.dxsummit.com
```

Passport to World Band Radio

Published by International Broadcasting Services, this reference guide is for shortwave broadcast listeners, now in its 20th edition.

Also, these Web sites can help you find out where to send your QSL card to confirm contacts with DX stations:

  ✔ **K4UTE:** www.nfdxa.com/K4UTE/K4UTE.HTML
  ✔ **IK3QAR:** www.ik3qar.it/manager/

Contesting

Contesting activity occurs every weekend, so if you want to join in, you'll need to have a good event calendar at your fingertips. As you learn to enjoy the sport of contesting, you'll want to learn about operating techniques and how the "regulars" make those big scores. Here are some resources that can speed your journey from "Little Pistol" to "Big Gun."

Contesting.com

Contesting.com is a Web portal that specializes in contesting and hosts several e-mail reflectors.

```
www.contesting.com
```

National Contest Journal

A bimonthly magazine published by the ARRL, National Contest Journal includes contest results and articles about contests and interviews. The NCJ also sponsors several popular contests.

```
www.ncjweb.com
```

Contester's Rate Sheet

This biweekly e-mail newsletter lists upcoming contests and contesting news, and covers technical topics and product release information. It's free to ARRL members.

```
www.arrl.org/contests/rate-sheet/about.html
```

And try these sites if you're looking for contest calendars with listings of upcoming events:

- **ARRL Contest Corral:** www.arrl.org/contests
- **By Bruce Horn WA7BNM:** www.hornucopia.com/contestcal
- **By Jan-Eric Rehn SM3CER:** www.sk3bg.se/contest
- **Mike Sivecic VK4DX:** www.vk4dx.net

Satellites

To contact any of the satellites whizzing around "up there," you need timely information about their status and orbitals. Getting started isn't nearly as difficult as you might think, especially if you can rely on the good how-to references listed here. The books are available through AMSAT or the ARRL.

Radio Amateur Satellite Corporation (AMSAT) Web site

This site provides the latest information on satellite status and links to useful information about satellite operating.

```
amsat.org/
```

Getting Started With Amateur Satellites, by G. Gould Smith WA4SXM

This book shows the beginner how to get started on satellites. It explains the necessary astronomical terminology, shows you how to locate the satellite and how to set up a satellite-capable station with inexpensive equipment.

The Radio Amateur's Satellite Handbook, by Martin Davidoff K2UBC

This handbook offers detailed information about all aspects of satellite operation.

The AMSAT-NA Digital Satellite Guide, published by AMSAT

This book provides step-by-step information about using the digital PACSATs.

Mobile operation

Do you prefer to do your hamming on the go? Here are some resources to answer all your questions about mobile operation:

Your Mobile Companion, by Roger Burch WF4N

An ARRL publication, this book provides an introduction to mobile operating, including how to set up the mobile station and operating guidelines.

The Mobile DXer, by Dave Mangels AC6WO

AC6WO's book shows how to do high-performance operating on the road, including tips on how to compete with the home stations. It is available from CQ Communications at www.cq-amateur-radio.com.

APRS Tracks, Maps and Mobiles - A Guide to the Automatic Position Reporting System, by Stan Horzepa WA1LOU

This guide, published by the ARRL, provides detailed information about how to set up your equipment to use APRS and use the Internet-based position-viewing software.

Technical References

Ham radio involves a lot of different technologies, and to perform at optimum level, you need to know as much as you can about all of them. The following sections detail some of the best resources for many different ham-related technologies.

General

Ham radio being the techie hobby that it is, you'll have more success if you can access technical references on a regular basis. Start with *The ARRL Handbook* and keep the Web links in your browser bookmark file.

The ARRL Handbook

Published by the ARRL, and now in its 81st edition, the handbook is a must for every shack. It's an encyclopedia of the technical aspects of amateur radio and includes numerous construction projects.

ARRL TIS Search

This site is an online reference covering many useful topics and containing links to articles from *QST* for ARRL members.

```
www.arrl.org/tis/tismenu.html
```

K1TTT Technical Reference

The K1TIT site offers numerous articles and links on the technical aspects of station building and design, with a heavy emphasis on antennas.

```
www.k1ttt.net/technote/techref.html
```

Digital Signal Processing Technology - Essentials of the Communications Revolution, by Doug Smith KF6DX

If you are interested in what's behind the DSP button on your radio, take a look at this book, which provides a detailed introduction to the topic and fully describes how DSP works.

Electronics

If you decide to go beyond the basic electronics knowledge required to get your license, there are many good texts. These three are good introductions to electronics and radio technology.

Understanding Basic Electronics, by Larry Wolfgang WR1B

This book helps you get started at the ground floor of electronics and learn about simple circuits and components such as the transistor and op amp.

RF Components and Circuits, by Joe Carr K4IPV

This book is a good introduction to RF circuit design and components. (All of Joe's books cover good introductory-level design and building techniques.)

33 Simple Weekend Projects for the Ham, the Student, and the Experimenter, by Dave Ingram K4TWJ

Here you find good starter projects in this book for useful gadgets in the ham shack and workbench.

Antennas

Hams are more likely to build antennas than any other piece of ham radio equipment. Thus, there is no shortage of ideas. These references contain many classic and innovative designs.

The ARRL Antenna Book

Published by the ARRL, this book is another ham radio classic, now in its 20th edition. It covers everything from basic antenna and transmission line theory to propagation and advanced antenna design. Useful construction projects round out every chapter.

"Classics" and "Compendium" series

The ARRL publishes both an *"Antenna Classics"* and *"Antenna Compendium"* book series, which consist primarily of construction project articles. The subjects include everything from simple wire antennas to complex arrays, microwave dishes, and transmission lines.

Backyard Antennas, by Peter Dodd G3LDO

Backyard Antennas is a great book on getting good performance from compact antennas that fit in limited space, perfect for the suburban or urban ham.

Cebik.com

L.B. Cebik W4RNL is a prolific author on antenna modeling and design. He has compiled a large number of articles on his Web site.

```
www.cebik.com
```

Qsl.net/wa1ion

Mark Connelley WA1ION specializes in receiving antennas and electronics for ham and SWL antennas. His Web site has many design articles and references to other useful sites, as well as a list of electronics design links.

```
www.qsl.net/wa1ion
```

Antenna Zoning for the Radio Amateur, by Fred Hopengarten K1VR

Here's a detailed guide to the process of working with local zoning regulations to obtain permits for amateur antennas and towers.

VHF/UHF/microwave

Operation above 50 MHz is one of the fastest-growing areas of ham radio with more and more excellent equipment and components becoming available every month. The following references help you assemble a working station and use it effectively.

VHF/UHF Handbook, edited by Dick Biddulph G8DPS

Published by the RSGB, the *VHF/UHF Handbook* is a comprehensive guide to setting up a station to go beyond the FM repeaters.

The ARRL UHF/Microwave Experimenter's Manual

Here you can find information and techniques for assembling and building equipment and antennas, and some guidance in understanding propagation.

50 MHz Propagation Logger

This site is a mini-portal aimed at the "Magic Band" operator. It features a real-time, worldwide propagation chat screen and lots of links to useful operating and propagation resources.

```
dxworld.com/50prop.html
```

144 MHz Propagation Logger

This site, a companion to the 50 MHz Propagation Logger Web site, handles 2-meter operating.

```
dxworld.com/144prop.html
```

Meteor Scatter

This page is for meteor scatter enthusiasts.

```
www.meteorscatter.net
```

Propagation

The science of propagation is truly fascinating and affects every ham. The more you know about it, the more interesting it becomes and the more success you'll have on the air. These references were selected to introduce the basics of radio propagation to the new ham.

The New Shortwave Propagation Handbook, by Jacobs, Cohen, and Rose

This handbook covers HF propagation from introductory levels to advanced topics. George Jacobs W3ASK wrote the CQ Magazine Propagation column for many years.

The Little Pistol's Guide to HF Propagation, by Robert R. Brown NM7M

This title, offered from Worldradio Books, explains HF propagation to the interested newcomer.

RSGB Propagation Page

Here you can find numerous links to information on HF and VHF/UHF propagation.

```
www.keele.ac.uk/depts/por/psc.htm
```

Spaceweather

This site offers news and information about the sun and the ionosphere, and includes links to real-time satellite photos and other information. You can also subscribe to e-mail alerts and data updates.

```
www.spaceweather.com
```

hfradio.org

Hfradio.org is a comprehensive site focused on HF propagation. It includes forecasts, discussions on ongoing events, and numerous charts and graphs showing historical propagation behavior.

```
www.hfradio.org/propagation.html
```

ARRL Propagation Bulletin

This weekly bulletin is about HF propagation, free to ARRL members.

```
www.arrl.org/w1aw/prop
```

And try these sites if you're looking for listings of propagation test beacons:

- ✔ **G3USF's 50 MHz Beacon List:** www.keele.ac.uk/depts/por/50.htm
- ✔ **10 Meter Beacon List:** www.ten-ten.org/beacons.html
- ✔ **Northern California DX Foundation's HF Beacon Network:** www.ncdxf.org/beacons.html

Amateur Magazines

The following magazines are the best of the general interest ham radio press. In addition to the ones listed in this section, every major organization likely has a membership magazine, as well.

QST

The ARRL's membership magazine has the greatest variety of technical, construction, and operating articles.

```
www.arrl.org/qst
```

CQ

Focused on general interest stories, product reviews, and columns, *CQ* also sponsors several major HF and VHF contests every year.

```
www.cq-amateur-radio.com
```

Worldradio

Worldradio specializes in columns and short general interest articles.

```
www.wr6wr.com
```

QEX

Short for *Calling All Experimenters, QEX* provides articles on state-of-the-art equipment and specialty articles of interest to the technically advanced ham.

CQ VHF Quarterly

This magazine is tailored to all types of operation above 50 MHz.

```
cq-amateur-radio.com/ourother.html
```

Vendors

The best way to become acquainted with the many ham radio vendors is to buy a copy of *CQ* or *Worldradio* (or check out *QST* from your local library) and scan the ads.

To look for a specific item, the ARRL Technical Information Service Web site (`www.arrl.org/tis/tisfind.html`) can help you find distributors or manufacturers of almost any ham radio-related item.

To find ham radio vendors on eBay, log on to `www.ebay.com`, and then browse through the Computers & Electronics and Radios: CB, Ham, and Shortwave categories. You can find other useful gear in the Software and Gadgets & Other Electronics categories. More test equipment is listed for sale in the Business & Industrial and Test Equipment categories.

And don't forget: Many of the ham radio Web portals also feature an online equipment swap-n-shop, including the ARRL, eHam.net, QRZ.com, and QSL.net portals I list in this appendix.

Index

• C •

• E •

• *N* •

• *O* •

FOR DUMMIES®

The easy way to get more done and have more fun

PERSONAL FINANCE & BUSINESS

0-7645-2431-3

0-7645-5331-3

0-7645-5307-0

Also available:

Accounting For Dummies
(0-7645-5314-3)

Business Plans Kit For
Dummies
(0-7645-5365-8)

Managing For Dummies
(1-5688-4858-7)

Mutual Funds For Dummies
(0-7645-5329-1)

QuickBooks All-in-One Desk
Reference For Dummies
(0-7645-1963-8)

Resumes For Dummies
(0-7645-5471-9)

Small Business Kit For
Dummies
(0-7645-5093-4)

Starting an eBay Business
For Dummies
(0-7645-1547-0)

Taxes For Dummies 2003
(0-7645-5475-1)

HOME, GARDEN, FOOD & WINE

0-7645-5295-3

0-7645-5130-2

0-7645-5250-3

Also available:

Bartending For Dummies
(0-7645-5051-9)

Christmas Cooking For
Dummies
(0-7645-5407-7)

Cookies For Dummies
(0-7645-5390-9)

Diabetes Cookbook For
Dummies
(0-7645-5230-9)

Grilling For Dummies
(0-7645-5076-4)

Home Maintenance For
Dummies
(0-7645-5215-5)

Slow Cookers For Dummies
(0-7645-5240-6)

Wine For Dummies
(0-7645-5114-0)

FITNESS, SPORTS, HOBBIES & PETS

0-7645-5167-1

0-7645-5146-9

0-7645-5106-X

Also available:

Cats For Dummies
(0-7645-5275-9)

Chess For Dummies
(0-7645-5003-9)

Dog Training For Dummies
(0-7645-5286-4)

Labrador Retrievers For
Dummies
(0-7645-5281-3)

Martial Arts For Dummies
(0-7645-5358-5)

Piano For Dummies
(0-7645-5105-1)

Pilates For Dummies
(0-7645-5397-6)

Power Yoga For Dummies
(0-7645-5342-9)

Puppies For Dummies
(0-7645-5255-4)

Quilting For Dummies
(0-7645-5118-3)

Rock Guitar For Dummies
(0-7645-5356-9)

Weight Training For Dummies
(0-7645-5168-X)

Available wherever books are sold.
Go to www.dummies.com or call 1-877-762-2974 to order direct

FOR DUMMIES®

A world of resources to help you grow

TRAVEL

0-7645-5453-0

0-7645-5438-7

0-7645-5444-1

Also available:

America's National Parks For Dummies
(0-7645-6204-5)

Caribbean For Dummies
(0-7645-5445-X)

Cruise Vacations For Dummies 2003
(0-7645-5459-X)

Europe For Dummies
(0-7645-5456-5)

Ireland For Dummies
(0-7645-6199-5)

France For Dummies
(0-7645-6292-4)

Las Vegas For Dummies
(0-7645-5448-4)

London For Dummies
(0-7645-5416-6)

Mexico's Beach Resorts For Dummies
(0-7645-6262-2)

Paris For Dummies
(0-7645-5494-8)

RV Vacations For Dummies
(0-7645-5443-3)

EDUCATION & TEST PREPARATION

0-7645-5194-9

0-7645-5325-9

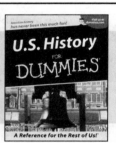

0-7645-5249-X

Also available:

The ACT For Dummies
(0-7645-5210-4)

Chemistry For Dummies
(0-7645-5430-1)

English Grammar For Dummies
(0-7645-5322-4)

French For Dummies
(0-7645-5193-0)

GMAT For Dummies
(0-7645-5251-1)

Inglés Para Dummies
(0-7645-5427-1)

Italian For Dummies
(0-7645-5196-5)

Research Papers For Dummies
(0-7645-5426-3)

SAT I For Dummies
(0-7645-5472-7)

U.S. History For Dummies
(0-7645-5249-X)

World History For Dummies
(0-7645-5242-2)

HEALTH, SELF-HELP & SPIRITUALITY

0-7645-5154-X

0-7645-5302-X

0-7645-5418-2

Also available:

The Bible For Dummies
(0-7645-5296-1)

Controlling Cholesterol For Dummies
(0-7645-5440-9)

Dating For Dummies
(0-7645-5072-1)

Dieting For Dummies
(0-7645-5126-4)

High Blood Pressure For Dummies
(0-7645-5424-7)

Judaism For Dummies
(0-7645-5299-6)

Menopause For Dummies
(0-7645-5458-1)

Nutrition For Dummies
(0-7645-5180-9)

Potty Training For Dummies
(0-7645-5417-4)

Pregnancy For Dummies
(0-7645-5074-8)

Rekindling Romance For Dummies
(0-7645-5303-8)

Religion For Dummies
(0-7645-5264-3)

Available wherever books are sold. Go to www.dummies.com or call 1-877-762-2974 to order direct